Welding

Welding

Techniques and rural practice

Peter Cryer

Jim Heather

INKATA PRESS

INKATA PRESS

A division of Butterworth-Heinemann Australia

Australia
Butterworth-Heinemann, 18 Salmon Street, Port Melbourne, 3207

Singapore
Butterworth-Heinemann Asia

United Kingdom
Butterworth-Heinemann Ltd, Oxford

USA
Butterworth-Heinemann, Newton

National Library of Australia Cataloguing-in-Publication entry

Cryer, Peter
Welding – Techniques and rural practice.

Includes index.
ISBN 0 7506 8921 8.

1. Welding – Technique. I. Heather, Jim. Title.
(Series: Practical farming).

671.52

Typeset by Ian MacArthur, Hornsby Heights, NSW.

Printed in Australia by Ligare Pty Ltd, Riverwood, NSW

Contents

Acknowledgments

The authors and the publisher wish to thank
The NSW TAFE Commission
CIGWELD – Comweld Group Pty Ltd
The Lincoln Electric Compnay (Australia) Pty Ltd
for their assistance in the compilation of this book.

Preface

*W*elding is designed to provide people working in rural enviroments with a broad range of welding skills. Coverage of the different welding and cutting processes is done in an easy-to-follow manner with consideration given to safety, equipment and application. To enable users to improve their skills with each process, practical problem solving is also included.

Repair and reclamation techniques are discussed and there is a chapter on basic marking out and development, which will enable farm welders to tackle the projects detailed in the final section.

This book is not intended to take the place of practice, but it will provide an excellent simple language guide to welding and serve as a handy reference when practical difficulties are encountered.

The TRACTAPAC makes manual metal arc welding possible wherever a tractor can go by using the tractor PTO to drive it (courtesy of the Lincoln Electric Company (Australia) Pty Ltd)

Introduction

Welding jobs in a rural environment are no different from other welding tasks. To repair a broken cross member on an agricultural implement requires the same detailed consideration as that required to repair a broken brace on a sand hopper support frame in urban industry.

However, the remoteness of the farm environment presents some special problems. Field work can often be many kilometres from a mains power outlet and a lack of ready access to stock materials makes repair by welding more common, and more viable, than removal and replacement of the broken part. Improvements in modern welding technology have made quality welding in the country much easier and cheaper than ever before.

Both gas metal arc welding (GMAW, MIG) machines and manual metal arc welding (MMAW, stick) machines are available for use directly from tractor power take–offs (PTOs), while smaller, fairly inexpensive but efficient machines can be obtained which need only a 240 volt outlet. These machines have almost eliminated the need for 415 volt, three phase power outlets. Tractor generated power machines usually have 240 volt outlets allowing hand tools (grinders, drills) to also be used in the field. When these machines are supported by oxyacetylene welding and cutting, rural workers have a very good portable workshop at their disposal.

Like all new technologies welding skills need to be learned and practised. Most modern metals are readily welded, but each one needs individual treatment. There are excellent welding courses conducted at most TAFE centres throughout Australia and to truly master all the aspects of welding these are highly recommended.

WARNING

Some welding operations are controlled by statutory bodies such as WorkCover NSW and welding on pressure vessels and pressure pipelines must be done or supervised by a person formally qualified in that field.

DON'T TAKE RISKS

If you are in any doubt about the safety of any welding situation seek qualified advice.

CHAPTER 1

Safety

Chapter contents

M ost of the machinery and equipment used in rural and industrial environments is potentially dangerous (tractors alone have been responsible for hundreds of deaths and serious injuries). Many of these acccidents have occurred when recommended safety precautions have been ignored. Welding equipment can also be dangerous, because welding produces heat, fumes and sparks, and it sometimes uses high voltage electricity, so it is also capable of causing death or injury. Awareness of the sources of these hazards is essential for the safe operation of all welding procedures.

This section looks at the precautions you need to follow to promote a safe working environment for yourself and those around you.

Electricity

The electricity needed to operate arc welding machines is supplied through mains power outlets or is produced by fuel driven generators.

Mains power supply

Mains power provides *alternating current* (a.c.) and is supplied at 240 volts at domestic installations and at 415 volts at industrial locations. Either voltage is capable of causing death or serious injury such as burns. Welding machines should be kept close to their power outlets and extension leads should not be used between the power outlet and the machine. If welding is required at a distance from the power outlet, then extended welding and return cables should be used.

Generators

Generators are usually driven by petrol or diesel motors and produce *direct current* (d.c.). The power take-off (pto) on tractors can be used to generate welding power, and welding machines with built-in generators are available for this purpose. The power produced by generators is also capable of giving a severe electric shock, and should be treated with the same respect as mains power electricity.

The welding circuit.

In all forms of arc welding an arc is established or struck betweed the job and an electrode. When this arc is established

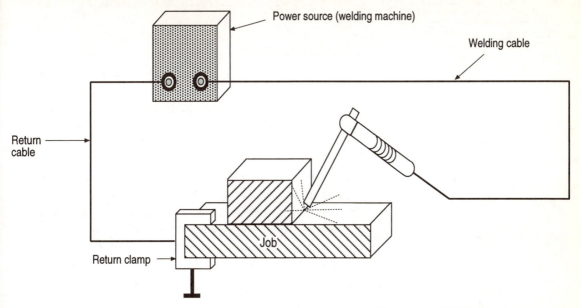

Figure 1.1 *Welding circuit*

an electric circuit is set up with electricity (power) flowing through it. *At no time should you become part of this circuit,* however, this can happen if welding is done on wet or damp sites, the equipment is faulty, is not grounded properly, or your protective clothing is unsuitable. Becoming part of the circuit is dangerous, but completely avoidable if safe working practices are followed (Figure 1.1).

Return clamp

The return clamp is attached to the return cable. When securely fitted to the job and the arc is struck it allows the welding current to flow in the curcuit. The clamp must be fitted to the job with clean metal-to-metal contact. Rust, paint, dirt and other non-metallic coatings should be removed from the clamping area by wire brushing or grinding. Failure to make clean contact between the job and the return clamp will create an unsafe condition and can severely affect the welding operation.

Changing electrodes

During manual metal arc welding it is necessary to replace electrodes as they are consumed. Always wear dry gloves when doing this. At no time should electrodes be held under the arm to change them. Even a small amount of sweat can

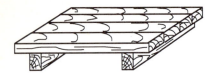

Figure 1.2 *Wooden duckboard*

create a pathway for the current and make you part of the circuit.

Insulation (underfoot)

Welding should always be carried out in dry surroundings.If the ground underfoot is wet or damp, wooden duckboards are needed to keep you clear of the moisture. This will prevent you from becoming part of the circuit (Figure 1.2).

High frequency attachments

Some welding machines have high frequency attachments which can be engaged to make it easier to strike and maintain the arc. Because these attachments operate at extremely high voltage they increase the chance of an electric shock. Protective gloves and boots must be worn for complete insulation when using this equipment, and at no time should such attachments be used in damp conditions.

Portable power tools

Most power tools, such as grinders and drills, are double insulated and have their outer parts constructed of non-conductive plastic. These machines offer good protection from electric shock, however, care should be taken not to damage leads by careless practice, especially when grinding. Older type power tools have cast metal casings and do not offer the same electrical safety as modern equipment. These machines should be regularly checked by a licensed electrician to ensure their safe working condition. Do not leave leads rolled up when in use and always switch off the power outlet before removing the three pin plug.

Heat and sparks

Fusion welding relies on the parts being joined to be melted at the joint area. This requires a great deal of heat and produces sparks as do most of the operations associated with welding.

Arc temperature

The arc temperature of most arc welding processes is around 6000°C– about four times the temperature of molten steel. This is the temperature at the actual point where welding takes place, and accordingly, is very localised, however, surrounding areas quickly become hot, because heat can be conducted a surprisingly long distance from the weld. Always wear gloves when welding and use tongs when handling small welded jobs. Small globules of molten metal are also expelled from the weld pool, which during welding appear as sparks. They can stay hot for a long time and can travel many metres from the job. These globules can be seen as spatter on welded work.

Oxyacetylene flame/oxy LPG (liquefied petroleum gas) flame

The oxyacetylene flame is the hottest flame used in industry. It produces a temperature of approximately 3100°C or about twice the melting point of steel. The oxy LPG flame produces a slightly lower temperature (about 2700°C).

Flame cutting

Flame cutting can be done using a variety of fuel gases, however oxyacetylene and oxy LPG are the most common. Apart from the high temperatures associated with these flames, flame cutting produces streams of molten iron oxide and you and those around you need protection, or very serious burning and eye damage can occur.

Just cut/welded work

After cutting or welding, metals can stay hot for a relatively long time. Just welded or cut work which is left unattended should be marked HOT and left where it cannot burn anything nearby or cause a fire.

Slag removal

The slag, which forms when manual metal arc welding, must be removed. It is hard and glass like, and is very hot on newly welded work. Eye protection must be worn, but care should also be exercised with areas of bare skin, because the hot slag can fly and 'stick', causing burns during its removal. You should wear gloves and a full-face head shield.

Grinding

As with all sparks, because grinding sparks are hot and capable of travelling long distances, you must take care where they are directed. They will permanently damage glass and high finished surfaces, such as paintwork on vehicles. Do *not* wear gloves when using pedestal or bench-mounted grinders, because the gloves can easily be 'dragged' onto the wheel and your hands can be severely damaged.

Butane gas cigarette lighters

These have no place in any environment producing heat and sparks, because they have been known to explode and cause death. Use lighting devices designed for welding workshops for your oxy torch and leave butane lighters at home.

Radiation

Intense visible light rays

All intensely heated objects give off radiation and most of this radiation is in the form of light and heat, which we can easily recognise.

Ultra-violet and infra-red radiation

Arc welding not only produces heat and intense light, but also produces other forms of radiation—ultra-violet and infra-red radiation. They are not visible and are similar to the radiation from the sun, which produces burning and eye damage. No sane person would stare at the sun without expecting to eventually go blind.

Radiation from welding is capable of temporary and permanent eye damage. The degree of damage is relative to the intensity, the length of time you were exposed to the arc and how close you were to it. Radiation will also burn and damage your skin.

To protect yourself from radiation wear your gloves, overalls and a head shield with an approved lens.

Fumes

Knowing what causes welding fumes can help you minimise them. Fumes are produced when the welding heat volatilises various substances in the weld area. Some of whese, such as fluxes, are required to be there, while others should be removed before welding. You should avoid being in the pathway of fumes and where possible use an extractor to remove them. Good ventilation is essential at all times.

Cleaning agents

Do not use degreasers containing trichlorethylene for cleaning as these can produce toxic fumes during welding.

Surface coatings

Oil, grease, paint and heavy rust all cause fuming and should be removed before welding.

Electrode coatings

The degree of danger from the fumes produced by electrode coatings varies. You should avoid breathing any of these fumes, and with hydrogen controlled and stainless steel electrodes, fume extractors are advisable.

Fluxes

Fluxes used in oxyacetylene welding and brazing are also a source of fumes.

Metallic coatings

Galvanising, chrome plating, nickel plating and cadmium plating produce dangerous fumes. These coatings can also reduce weld quality. They should be removed by grinding before welding commences.

Metal additives

When metals are heated to high temperatures they form metal oxides. Some of these oxides are toxic; chromium, nickel, zinc, cadmium and beryllium oxides are particularly dangerous.Effective fume extractors or respirators should be worn when working on alloys containing these metals.

Personal protection

Protective equipment is especially made to protect you from welding hazards. You must use this gear at all times while you are welding; even doing small jobs without protection, because you are in a hurry or it is just too much trouble to use it, is a sure-fire way to disaster. It is too late to reach for your gear after an accident, so always have the recommended protective equipment on hand and use it.

Your basic protective equipment should consist of the following items.

Protective clothing

Full overalls or long sleeved shirt and long trousers. These should be made of work quality cotton or wool. Some materials used to make clothing are treated with a protective solution which is flammable, so new clothing should be washed before it is worn in a welding workshop. Do not wear torn, frayed or greasy clothing, and do not leave matches and other flammables in your pockets.

Footwear

The ideal welding footwear is solid leather shoes or boots with steel toe caps and rubber soles. The elastic sided style is probably the most suitable, because they reduce the chance of sparks entering via the lacing gap.

Never weld in open topped or synthetic footwear.

Gloves

These must be made of leather and should be the gauntlet style, covering your forearms.

Welding shields

These are available as a hand-held shield or a head shield which leaves both hands free. They are needed to filter out dangerous radiation and reduce glare. Only shields with a lens designed for arc welding are suitable; your shield should have a symbol indicating that it meets the relevant standard.

Chipped or cracked lenses should be replaced befor welding and the lens and its protective plastic clear lens should be cleaned regularly and replaced if they become scratched or pitted. A shade 10, 11,or 12 lens is suitable for most welding. The higher the number the darker the lens.

Oxy goggles

These are essential for oxyacetylene welding, braze welding, brazing and oxy-fuel gas cutting. Like the arc welding lens these are approved and carry identifying marks. A shade 5 lens will satisfy most welding and cutting operations. Serious eye damage can result if non-approved lenses such as sunglasses are used.

Safety glasses

Clear lenses must be worn when you are not welding, because you must protect yourself from grinding and slag

Figure 1.3 *Welding shield and gloves (courtesy of CIGWELD)*

removal. These safety glasses must be made of shatterproof glass or clear plastic and kept clean. Some welding head shields and oxy goggles have flip-up lens which leave your eyes protected when not welding. These are all right for short periods between welding operations, but can become stuffy and uncomfortable when worn for long periods.

Always have your safety glasses handy and never grind or de-slag without your eyes being protected.

Figure 1.4 *Metal-framed safety spectacles (courtesy of CIGWELD)*

Ear muffs and ear plugs

If you expose yourself to loud noises for long periods you will go deaf. Hearing loss is completely avoidable. When you are working with noisy machinery, such as grinders or in noisy environments, you should wear ear protection. Make sure you know how to fit and use your ear plugs or ear muffs; your welding equipment supplier should be able to advise you on this.

When does a noise become dangerous? A good guide is that if you need to raise your voice above normal conversation levels to be heard or understood, then you need to protect your hearing.

Protection of others

When using welding or cutting equipment you have a legal obligation to protect people around you. Most people will not be fully aware of the dangers of welding and may in fact be tempted to watch you while you work.

Children should be kept well clear of all welding and cutting operations. Some things you can use to protect others while you are working are as follows.

Figure 1.5 *Flash screen (courtesy of CIGWELD)*

Screens

The welding area can be partitioned off from other areas by portable screens made from non-flammable material. These can be moved to accommodate the size of the job or arranged to shield the welding area on larger jobs. Being portable, they can be easily stored when not required.

Signs

When welding in a location where the public may be, warning signs should be used. These signs should be displayed at a barrier at least ten metres from where the welding is taking place.

Observers

Observers can be used to keep a watch for hazards and to warn people of possible dangers.

Confined spaces

Description/definition

Sometimes we are required to carry out welding or cutting operations in areas with restricted ventilation or limited entry and exit options. These areas are classed as confined spaces by the regulatory authorities and special conditions apply to them.

Examples of confined spaces

- silos
- tanks
- pipes

Inspection

Before any work is done in a confined space an assessment of the area should be made to determine how the ventilation can be improved and how, in an emergency, evacuation or rescue from the area can be conducted.

Permits

On some work sites you may require an entry permit before you begin working in the area. The relevant authority in your area, (Workcover Authority of NSW), can advise you of these requirements.

Ventilation

As with all welding and cutting you must provide adequate ventilation when working in a confined space. Forced ventilation using air only (not oxygen) will help, however, removal of fumes from the source using extractors is advisable. Respirators with dust and fume filters should be used, while respirators with outside air supply may be needed on some jobs.

Access and egress

Never enter a confined space without first establishing how you will get out. Secure ladders and leave them in place. Leave all access doorways or hatches open and wear a body harness with a rope attached. Keep the pathways to your exits clear of tools and equipment.

Observer

A responsible person should be present to observe you while you are in a confined space. Your observer must be able to see you at all times and be prepared to act quickly in the case of an emergency.

Precautions

When working in a confined space

- Provide yourself with good ventilation or use a respirator.
- Have an observer.
- Have appropriate fire extinguishers on hand.
- Wear protective clothing and use wooden duck boards when working in a damp area or in metal tanks.
- Ensure all cutting and welding equipment is in good order – no leaks or damaged leads.
- Keep cutting equipment outside the container or space. Have your assistant light the torch outside and hand it in to you to use. To make sure that no explosive gases build up due to leaks in the equipment, the torch should not be extinguished inside the work area.
- Keep access to your exit clear of tools and equipment.
- Allow your observer to keep you in view at all times.
- Wear a rescue harness attached to a rope held by your observer.
- When it is necessary obtain an entry permit from your statutory authority.
- Use pneumatic or 24 volt hand tools.

Hazardous locations

Description/definition

Hazardous locations are sites where there is a high risk of fire or explosion. Before welding or cutting in these locations strict safety precautions need to be adopted.

Examples of hazardous locations:

- all confined spaces
- oil refineries
- service stations
- saw mills
- grain silos
- any location where flammable or explosive substances are on site, manufactured or have been stored.

Inspections

Inspection of the site by a qualified person is essential. This person will usually determine any additional precautions you may need to adopt before and during your work at that site. These precautions will be given to you in writing and you will need to follow them closely.

Permits

You must obtain a **permit to weld and cut** before working at a hazardous location. This will be issued by a person deemed to be qualified to do so by the statutory authority in your area. (In NSW this authority is Workcover.) If you have any doubts about welding at a particular location, you should not proceed until you are sure it is safe to do so. If you need further advice in this matter contact your authority. A permit to cut and weld will include:

- The exact location on the site where the cutting and welding is to be done. This will sometimes include a brief description of the task.
- The date and the time allowed for the work to be done. Permits can only be issued for one day at a time. A new permit is required each day.
- Any special precautions you may need to follow such as clearing or wetting down the area.
- The name of the person the permit is issued to and the name and status of the person issuing the permit.

Observer

As with confined spaces a responsible person has to be on hand to watch for any signs of danger and to assist the welder.

Precautions

Standard precautions when working at a hazardous location are as follows:

- Have the area inspected and obtain a permit to cut and weld.
- Have an observer with you.
- Clear the area of all flammable material for a radius of at least ten metres and cover or wet down wooden floors.
- Ensure all equipment is in good condition.
- Keep suitable fire extinguishers or other fire fighting equipment nearby and know how to use it.
- Observe all special precautions stated on your permit to cut and weld.
- Remain on site for at least one hour after you have finished cutting and welding.

Containers and drums

The heat from cutting or welding will quickly volatilise flammable residues in a container, making an explosion possible. Fuel containers (petrol, dieseline, kero and mineral turpentine) are obviously dangerous to cut or weld, but other containers which appear safe may also be hazardous. Exercise caution when cutting drums which have held insecticides or fertilisers, because some of these may produce explosive conditions and need to be thoroughly cleaned first. Merely flushing with cold water is not good enough and more stringent cleaning procedures must be used.

Cleaning

All residue should be removed from the container by draining or if possible scraping. The container should then be washed with detergent and boiling water; this should be repeated until the container is clean. It should be filled with water to just below the area to be cut or welded. Provided the air space at the top of the container is vented to the atmosphere work can now commence.

Steaming

Ideally, containers which have held flammable substances should be steam cleaned. This can be done using the portable

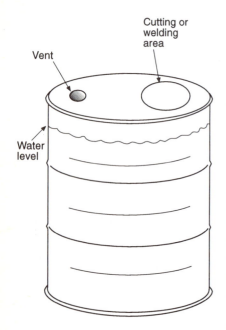

Figure 1.6 *A container prepared for welding*

Labels on figure: Vent, Cutting or welding area, Water level equipment used to steam clean motor vehicle engines. Steam cleaning should take place for at least half an hour or until the container is hot to touch.

Wear suitable clothing, gloves and face protection when using steam cleaning equipment.

Purging
Containers can be filled with a non-flammable gas such as carbon dioxide after cleaning, to further reduce the chance of explosion. Displacing the air in the container removes the oxygen needed to support the ignition of any flammable residue.

Working on ladders, scaffolding or other elevated positions

You have a responsibility to protect other people when you are working in an elevated position. Scaffolding over three metres high must be erected by a licensed scaffolder. Licensing is also required to operate hoists which incorporate working platforms or buckets, such as the *cherry-picker* type.

Observer

An assistant stationed on the ground is a must. This person can:

• Assist in setting up and dismantling.
• Ensure that there is no danger from falling sparks or tools.
• Warn others who may walk into the area.
• Help secure equipment and adjust welding settings when required.
• Load or tie welding requirements onto any lifting device for the operator.

Signs

When working in a public place or where there are others around you, you should rope off the area below you and display warning signs.

Kickboards

When working on platforms, kickboards should be in place to prevent accidentally dropped hand tools falling off the edge of the platform.

DANGER
WORKERS OVERHEAD
DO NOT ENTER

Figure 1.7 *Typical warning sign*

Securing equipment

Ladders should be tied into position; if this is not possible an assistant must be on hand to hold the ladder while it is being used. Welding cables, power leads and cutting equipment hoses should be tied securely to the scaffold or platform to prevent them falling, and to reduce the weight the operator needs to hold when using them.

A rope can also be tied to the work platform to raise and lower additional tools and equipment.

Lifting gas cylinders

These should be raised or lowered using a cylinder cradle and at no time should cylinders be lifted by attaching ropes to the brass spigot at the top of the cylinder.

Wire cables and rope

Where wire slings, cables or ropes are used for lifting or securing scaffolding or equipment, care should be exercised so that they are not burnt or arced across by cutting or welding equipment. Damaged ropes or slings should be replaced immediately.

Head protection

Hard-hat head protection is essential for those working on elevated positions, and those below must also wear hard hats. A welding head shield is available which incorporates a hard hat.

Figure 1.8 *Platform with surrounds*

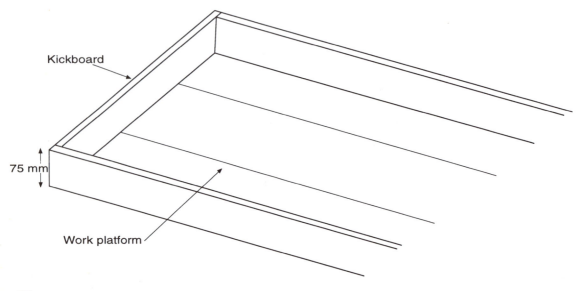

Kickboard

75 mm

Work platform

Weld joint terminology

A common welding language is needed so that we can understand each other when discussing welding. Without a standard terminology, welding instructions would be very difficult to follow, and confusion and unsafe practices could easily occur. We need to know the different types of welds, their parts and the terms used to identify unacceptable weld conditions.

The welding terminology covered in this chapter is similar to that used in literature produced by the Standards Association of Australia and counterpart organisations overseas.

Fillet welds

These welds are generally used to join surfaces which meet at an angle to each other. A fillet weld can be seen as a triangular weld external to the parts being joined. Before parts can be successfully fillet welded they should be thoroughly cleaned so that all oil, grease, paint, galvanising or other coatings are removed.

Types of fillet welds

Corner fillet weld

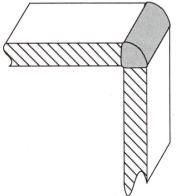

Figure 2.1 *A corner fillet weld joining two plates*

This type of weld is used to join plates when their ends meet at an angle to each other (usually 90°). The corners of rectangular tanks can be joined using corner fillet welds (Figure 2.1).

Lap fillet

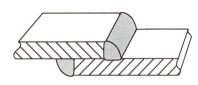

Figure 2.2 *Two plates joined by lap fillet welds*

These joints are used to join plates together in a continuous line. One plate overlaps the end of another plate and a fillet weld is done on each side. This type of weld joint is usually used for thin metal,because it is easier and often stronger than placing the parts end-to-end. It can be used on thicker plates if the step it creates is acceptable (Figure 2.2).

Tee fillet

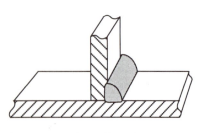

Figure 2.3 *A tee fillet weld*

These welds are common in metal structures. When the end of one plate meets the surface of another tee, fillets are often used (Figure 2,3).

Plug fillet

These welds are used to join two flat surfaces together. When an angle bar or channel frame is covered with plate, holes are drilled through the plate, allowing fillet welds around the circumference of the hole to be used to join the plate to the frame. These welds can also be used to join two plate surfaces together to produce a thicker plate. Completely filling the holes in plug welding is not necessary; the fillet will produce a strong joint (Figure 2.4).

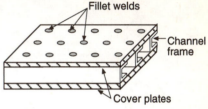

Figure 2.4 *A typical application of plug fillet welds*

Slot fillet

These welds are similar to plug fillet welds, however, instead of holes, round ended slots are made (generally by flame cutting) (Figure 2.5).

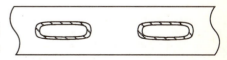

Figure 2.5 *Slot fillet welds*

Parts of a fillet weld

Parent metal

Parent metal is the metal being welded.

Weld metal

Weld metal is all the metal melted during the welding. This will usually be a mixture of parent metal and added metal (filler wire or electrode).

Weld face

Weld face is the outer surface of a weld run (Figure 2.6).

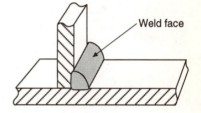

Figure 2.6 *Weld face*

Weld face contour

The volume of deposited metal will influence the weld face contour. The shape of the weld face will be concave, mitre or convex (Figure 2.7).

Figure 2.7 *Weld face contours*

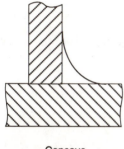

Concave

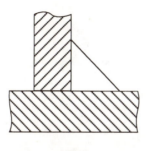

Mitre

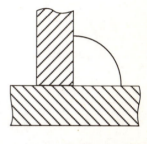

Convex

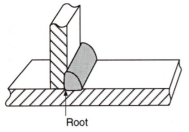

Figure 2.8 *Root*

Root

The root is the innermost part of a weld directly opposite the face (Figure 2.8).

Toe

The toe is the junction of the weld face and the parent metal,or in multi-run welds, a weld toe forms the junction of the weld faces (Figure 2.9 and 2.10).

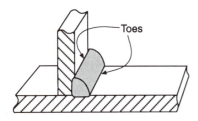

Figure 2.9 *Toes*

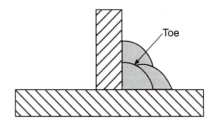

Figure 2.10 *A toe at the junction of weld faces in a multi-run fillet weld*

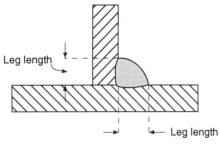

Figure 2.11 *Leg length*

Leg length

The leg length is the distance from the root of the weld to its toe (Figure 2.11).

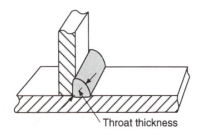

Figure 2.12 *Throat thickness*

Throat thickness

The throat thickness is the distance from the root to the face of a weld (Figure 2.12).

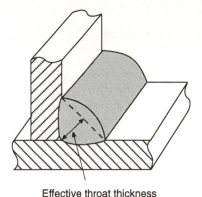
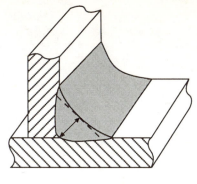

Effective throat thickness

Figure 2.13 *Effective throat thickness*

Effective throat thickness

The effective throat thickness is the distance from the root to the centre of the hypotenuse (long side) of the largest isosceles triangle which can be drawn in the sectional shape of the weld. In a convex fillet this is the distance from the root to the centre of a line joining the weld toes. In concave (and mitre) fillet welds, the effective throat thickness is the same as the weld's throat thickness (Figure 2.13).

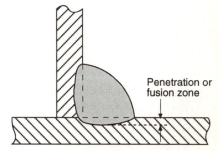

Penetration or fusion zone

Figure 2.14 *Penetration or fusion zone*

Penetration

Penetration is the depth the weld metal extends into the parent metal. This is also referred to as the fusion zone of a weld (Figure 2.14).

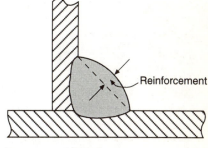

Reinforcement

Figure 2.15 *Reinforcement*

Reinforcement

The reinforcement is the distance the weld metal extends beyond a line joining the weld's toes (Figure 2.15).

Heat affected zone (HAZ)

The heat affected zone is the parent metal adjacent to the fusion zone. This metal has not been melted, but has been altered metallurgically by the welding heat (Figure 2.16).

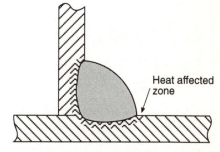

Heat affected zone

Figure 2.16 *Heat affected zone*

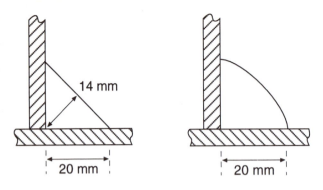
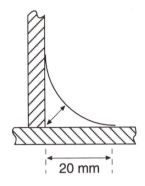

Acceptable 20 mm fillet welds

Undersized throat
for a 20 mm fillet weld

Figure 2.17 *Size of fillet welds*

Size of a fillet weld

A weld identified as a 10 mm fillet weld will have a 10 mm leg length; however, two measurements are used to determine the size of a fillet weld. These are leg length and throat thickness. The throat thickness in a fillet weld should be a minimum of 0.7 times its leg length. That is, a 20 mm fillet weld must have a minimum throat thickness of 14 mm (Figure 2.17).

Butt welds

The weld metal in butt welds must extend for the full thickness of the plate, and special edge preparations are needed on plate over 3mm thick to achieve this. The different types of butt welds are named after the shape of the edge preparation used. The geometry and methods of edge preparation are discussed in Chapter 3.

Types of butt joints

Close square butt joint

Figure 2.18 *Close square butt joint*

The close square butt joint is used to join metal up to 1.5 mm thick (Figure 2.18).

Open square butt joint

Figure 2.19 *Open square butt joint*

The open square butt joint is used to join metal up to 3mm thick (Figure 2.19).

V butt joint

To make a V butt joint the edges of the plates are bevelled before welding. This may be from one side (single V) or from both sides (double V) depending on the plate thickness (Figure 2.20).

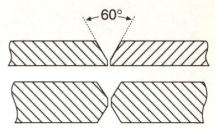

Figure 2.20 *V butt joint*

U butt joint

As with V butt joints, U butt joints may be single U or double U (Figure 2.21).

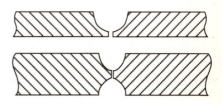

Bevel butt joints

Bevel butt joints are used where the end of a plate joins onto a flat surface. This is similar to a tee fillet, but the edge of one plate is prepared to ensure a full penetration weld (Figure 2.22).

Figure 2.21 *U butt joints*

Single bevel butt joint   Double bevel butt joint

Figure 2.22 *Single and double bevel butt joints*

Parts of a butt weld

Parent metal

The parent metal is the metal being welded (Figure 2.23).

Parent metal

Figure 2.23 *Parent metal*

Weld metal

The weld metal is all the metal melted during the making of a weld (Figure 2.24).

Weld metal

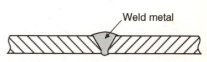

Figure 2.24 *Weld metal*

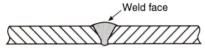

Figure 2.25 *Weld face*

Weld face
The weld face is the outer surface of a weld run (Figure 2.25).

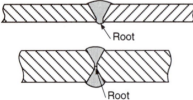

Figure 2.26 *Root*

Root
The root is the innermost part of a weld directly opposite its face (Figure 2.26).

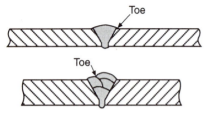

Figure 2.27 *Toe*

Toe
The toe is the junction of the weld face and the parent metal, or in multi-run welds the junction of the weld face and existing weld metal (Figure 2.27).

Figure 2.28 *Root penetration*

Root penetration
The root penetration is the extent to which the weld metal fills the root of the joint or extends past it (Figure 2.28).

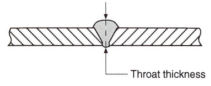

Figure 2.29 *Throat thickness*

Throat thickness
The throat thickness is the distance from the root of a weld to the face of the weld. In full penetration butt welds, the throat thickness is the same as the plate thickness (Figure 2.29).

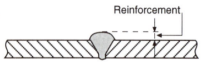

Figure 2.30 *Reinforcement*

Reinforcement
The reinforcement is the distance the face of the weld extends beyond the surface of the parent metal (Figure 2.30).

Fusion zone

The fusion zone is the area where the weld metal has penetrated the parent metal (Figure 2.31).

Figure 2.31 *Fusion zone*

Heat affected zone (HAZ)

The heat affected zone is the parent metal adjacent to the fusion zone. This metal has not been melted but has been metallurgically altered by the welding heat (Figure 2.32).

Figure 2.32 *Heat affected zone*

Size of a butt weld

In general the throat thickness of a butt weld should be equal to the plate thickness. The joint should have full penetration and be completely filled (Figure 2.33).

Figure 2.33 *Weld size*

Welding positions

The welding position is determined by where the welding operator is positioned in relation to the joint and the joint axis. Broadly these positions are as follows.

Flat position

Flat position welds are done from above the joint while its axis is approximately horizontal (Figure 2.34).

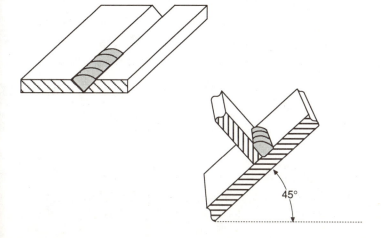

Figure 2.34 *Flat position*

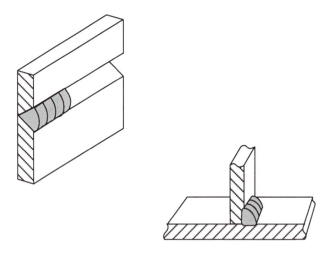

Figure 2.35 *Horizontal position*

Horizontal position

Horizontal position welding is done from the side (or in front) of the joint while its axis is approximately horizontal (Figure 2.35).

Vertical position

Vertical position welding is done from the side or the front of the joint, while its axis is approximately vertical (Figure 2.36).

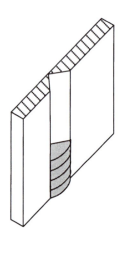
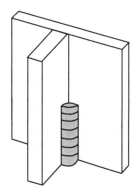

Figure 2.36 *Vertical position*

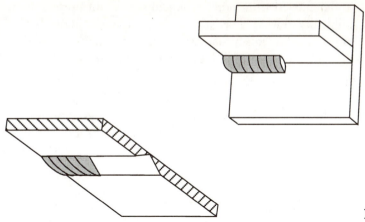

Figure 2.37 *Overhead position*

Overhead position

Overhead position welding is done from below the joint while its axis is approximately horizontal (Figure 2.37).

Pipe joint positions

Joints in pipes where the pipe axis is vertical are the same as those in plate and are identified as a horizontal pipe joint (Figure 2.38).

When pipes which cannot be rotated (fixed position) are being butt welded, the welding position changes as the welding progresses. These joints are identified as:

Pipe axis horizontal – fixed position. The welding in this joint changes from overhead to vertical to flat, when the weld is started from underneath the joint (Figures 2.39 and 2.40).

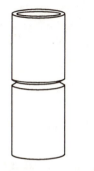

Figure 2.38 *Horizontal position pipe joint*

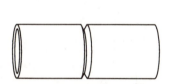

Figure 2.39 *Pipe axis horizontal – fixed position*

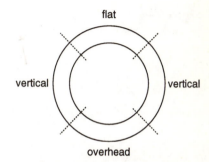

Figure 2.40 *Cross section of a pipe indicating approximate weld positions*

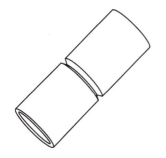

Figure 2.41 *Pipe at 45° angle*

Pipe axis inclined – fixed position. This pipe position is differentiated from other pipe joints as it can present some difficulty to weld. The parts of the joint do not fit neatly into any of the previously discussed positions, and during welding a higher degree of skill is generally needed (Figure 2.41).

It is important for welders to be able to weld efficiently in all welding positions, however, where possible, welding should be done in the flat position. In this position good welds are usually easier to achieve and it is generally more comfortable for the operator.

Weld faults

Weld faults are imperfections in the welded joint which *reduce its strength*. Slight imperfections which do not significantly affect the serviceability of a weld can occur and are not classed as faults. With practice weld faults can be avoided. (Weld fault prevention is dealt with as part of the discussion on welding processes.)

Figure 2.42 *Misalignment in a butt joint*

Misalignment

Misalignment occurs at a joint where the parts have not been assembled correctly before welding, or have moved out of alignment during welding. Misalignment can increase the stress to the joint area and reduce its strength. This is different from distortion which is discussed in the next chapter (Figure 2.42).

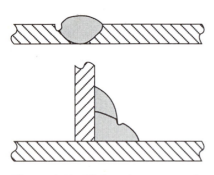

Figure 2.43 *Undercut in parent metal* (above) *and in weld metal* (below)

Undercut

Undercut is a groove melted in the parent metal (or in an existing weld metal) at the toe of a weld. This fault reduces the thickness of the metal being welded and provides a fault line where stresses can concentrate (Figure 2.43).

Overroll

If a weld pool is allowed to become too large the molten weld metal may collapse and run onto unmelted parent metal. This provides a weld where little or no fusion has occurred at the toe of the weld. Under stress this area can readily tear (Figure 2.44).

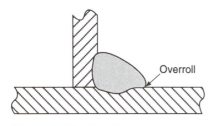

Figure 2.44 *Overroll*

Inclusions

Any foreign matter trapped in a weld is referred to as an inclusion. This can result from:

- unclean surfaces (paint, grease or rust)
- metal coatings, such as galvanising
- failing to remove the slag from previous runs.

Often the inclusions are not visible on the surface of the weld and can only be detected by destructive or non-destructive testing.

Porosity

Porosity is the pores (bubbles) left by gas which has been unable to escape from the weld. These pores take the place of sound weld metal and reduce the strength of the weld (Figure 2.45).

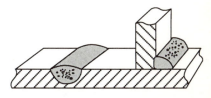

Figure 2.45 *Internal porosity* (left) *and surface porosity* (right)

Lack of root penetration

The failure of the weld to penetrate the parent metal can result in the metal being only partially welded. This will reduce the strength of the weld (Figure 2.46).

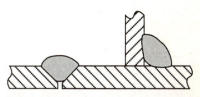

Figure 2.46 *Lack of root penetration*

Lack of fusion

When the parent metal and the filler metal are not melted sufficiently, a proper union does not occur. In extreme cases no true weld has taken place, however, generally there is only lack of fusion in part of the joint (Figure 2.47).

Figure 2.47 *Lack of side wall fusion in a butt joint*

Undersized and oversized welds

The thickness of the metal being welded will usually determine the weld size. Fillet welds in 10 mm thick plate will commonly have 10 mm leg lengths. When plates of different thicknesses are welded together the weld size is determined by the thinnest section. Oversized welds will set up stress points because they create a sharper angle between the parent metal and the weld metal. Under stress, tearing can occur in the parent metal at the toe of the weld. Sometimes a smaller weld is acceptable, however the weld must always be equal to the weld size specified (Figure 2.48).

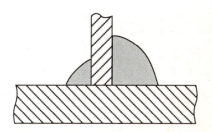

Figure 2.48 *Correct weld size* (left) *and oversized weld* (right)

Cracking

Fractures in parent metal or weld metal are referred to as cracks. Most cracking in metal as a result of welding can be

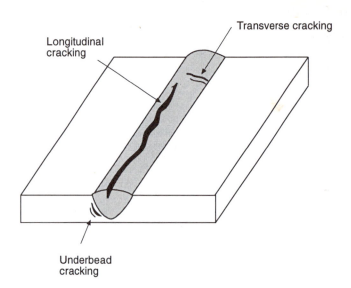

Figure 2.49 *Some types of weld cracks*

traced to the welding procedure. Either the procedure is unsuitable or it has not been adhered to by the welder. Cracks are generally described by their location in the weld or weld area, and can be in the weld metal or the parent metal. Some of these are:

Crater cracking
Cracking at the end of a weld run where the arc has been broken.

Longitudinal cracking
Cracking along or parallel to the axis of the weld.

Transverse cracking
Cracking across the axis of the weld.

Toe cracking
Cracking in the parent metal at the toe of the weld.

Underbead cracking
Cracking in the heat affected zone of the weld.

Excessive splatter
When the droplets of weld metal expelled from the weld pool become excessive they affect the appearance of the weld. You should try to reduce splatter levels, especially on stainless steels where they can reduce its corrosive resistance.

Edge preparation and minimising distortion

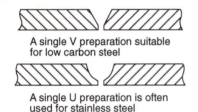

A single V preparation suitable for low carbon steel

A single U preparation is often used for stainless steel

Figure 3.1 *Simple edge preparations*

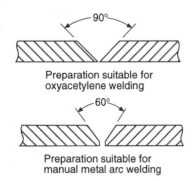

Preparation suitable for oxyacetylene welding

Preparation suitable for manual metal arc welding

Figure 3.2 *Joint angles*

Edge preparation is the work needed to prepare the edges of the items to be joined, and is essential for most welded joints. The type of edge preparation needed will be influenced by several factors.

- As the thickness of the metal increases so too will the need to prepare the metal from both sides.

- The edge preparation selected for one type of metal may not be suitable for another. Some 'easy-to-do' preparation can create too much stress in joints on harder or hardenable metals, making it necessary to use an alternative preparation which causes less stress (Figure 3.1).

- Some welding processes require more room to manipulate, influencing the geometry of the joint. The joints for oxy acetylene welding will usually require a 90° included angle, while those for manual metal arc welding will require a 60° included angle (Figure 3.2).

- When working from specifications the type of edge preparation will be stipulated by the engineer or customer.

Reasons for edge preparation

Edge preparation is needed to:

- enable side-wall fusion for the full thickness of the material
- allow satisfactory root penetration
- help control distortion (distortion is dealt with at the end of this chapter)
- produce a stronger weld, with less internal stress.

Cleaning

The parts which make up the weld joints must be clean. This requires the removal of grease, oil, paint and any other coatings from the area to be welded. Cleaning is important for both butt and fillet welds.

Irregularities

Prepared edges should be clean and free of irregularities. Where necessary flame-cut portions should be ground smooth.

Some terms related to edge preparations for butt welds

Figure 3.3 *Bevel*

Bevel

The bevel is the angle of a prepared edge forming part of a joint (Figure 3.3).

Figure 3.4 *Root face*

Root face

The root face is the part of the prepared edge of a joint that is not bevelled (Figure 3.4).

Root gap

The root gap is the distance between the parts at the root of the weld when they are set up ready to join. This gap is necessary to allow root penetration when welding (Figure 3.5).

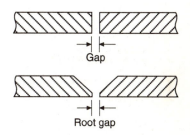

Figure 3.5 *Root gaps*

Included angle

The included angle is the angle between the prepared edges forming a joint (Figure 3.6).

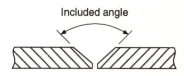

Figure 3.6 *Included angle*

Root radius

The root radius is the radius of the curve forming part of the preparation of a U joint (Figure 3.7).

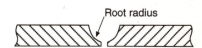

Figure 3.7 *Root radius*

Edge preparation methods

Flame cutting

Flame cutting is by far the commonest method used to prepare plate edges for welding. Machine flame cutting, such as with straight-line cutters, produce edges which require little more than wire brushing before welding, however, root faces may need to be added by filing or grinding. Hand held flame cutting edges will need to be dressed before welding.

Flame gouging

Flame gouging can produce the U joint edges. This technique is an adaptation of oxyfuel gas cutting and requires a great deal of skill. (see flame gouging in Chapter 6.)

Grinding

Grinding is needed to finalise the preparation of some flame cut edges. Grinding will be more suitable than flame cutting when only a small volume of metal is to be removed for joints on thin plate

Machining

Machining is very expensive, but when U joints have to be used, they are best prepared by milling machines. A lathe can be used to machine bevels on the ends of short lengths of pipe.

Distortion

During welding distortion occurs because the expansion and contraction of the metals being joined is uneven or restricted.

Expansion and contraction.

When a piece of metal is evenly heated it expands (gets bigger) in all directions, so after heating it will be longer, wider and thicker than it was before. The higher the temperature the bigger it becomes. If allowed to cool without restriction, the piece of metal will return to its original size and shape, because contraction (becoming smaller) also takes place in all directions. When only part of a piece of metal is heated – as is the case during welding – expansion is restricted by the unheated parts. These cooler areas prevent the hot metal from moving freely, thus the expanding area moves mostly in the directions with the least resistance to its expansion. Expansion has now occurred unevenly.

Contraction will still take place evenly and in all directions and when it is cool the metal will have a different shape from what it had before heating took place.

Different metals expand by different amounts when their temperature is raised to the same level. For example, aluminium expands about twice as much as steel when exposed to the same temperature increase. Copper and austenitic stainless steel also have greater expansion rates than steel. The greater the expansion rate of a metal the more likely it is to distort during welding.

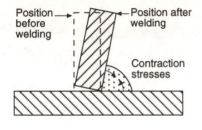
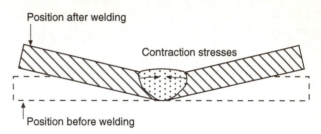

Figure 3.8 *Angular distortion*

Types of Distortion

Angular distortion

Angular distortion is distortion which results in a change of the angle between the parts being welded (Figure 3.8).

Longitudinal distortion

Longitudinal distortion occurs along the line of the weld. This type of distortion is often a problem when butt welds in plate are made from one side only . Because of uneven expansion and contraction, after welding, the length of the weld face is shorter than the root of the weld, and the parts will be curved (Figure 3.9)

Figure 3.9 *Longitudinal distortion*

Transverse distortion

Transverse distortion occurs across the line of the weld. When plates are butt welded without sufficient tacking or clamping , a scissor effect takes place after a short portion of the joint has been welded. Under these conditions the contraction stresses make it impossible to maintain the desired alignment, and the weld will need to be removed and a more adequate tacking or clamping procedure adopted (Figure 3.10).

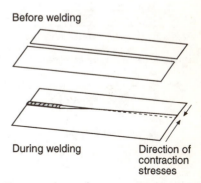

Figure 3.10 *Transverse distortion*

Minimising distortion

Before welding

Adopting procedures which will prevent distortion form happening is good work practice. You will save a lot of time by not having distortion problems to rectify later. Your job will look better and serve its purpose better. Some things you can do which will help prevent distortion are as follows.

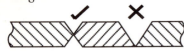

Figure 3.11 *Use double joints where practical*

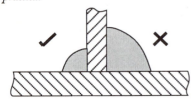

Figure 3.12 *Don't overweld*

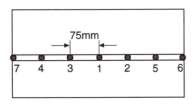

Figure 3.13 *A typical tacking procedure for a butt joint in 3 mm thick plate*

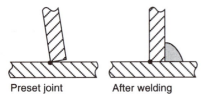

Preset joint   After welding

Figure 3.14 *Allowing for movement*

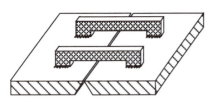

Figure 3.15 *'Strongbacks' being used to jig a butt joint*

Design

Design your jobs so that there are no highly stressed areas.

- Use double V or double U joints rather than single ones (Figure 3.11).
- Don't over weld. A 20mm fillet on 10mm thick plate will increase the contraction stresses at the joint (Figure 3.12).

Process selection

The less heat you put into the joint the less it expands and contracts, and the less it distorts. While we must melt the metal to weld it, the heat input of individual welding processes varies. The heat input of manual metal arc welding (MMAW) is higher than that of gas metal arc welding (GMAW) while oxyacetylene welding (OAW) has a higher heat input than both of these.

Tacking

Ensure that your tacking procedure is effective. Watch the order in which you do the tacks, their size and their frequency. Metals with high expansion rates will need larger tacks or at more frequent intervals (Figure 3.13).

Presetting

If you estimate the movement that is likely to take place during welding and allow for it, it is possible to achieve the required alignment and not restrict the movement of the parts. This is done by tacking the parts together to form a slightly greater angle than the finished job will have (Figure 3.14).

Clamping or jigging

By completely preventing the parts from moving during welding, distortion can be minimised. However, rigid clamping and jigging can increase the chance of weld cracking especially in harder or hardenable metals (Figure 3.15).

During welding

The best way to minimise distortion during welding is to use a weld sequence which prevents the contraction stresses from concentrating in one area or direction. Sequence welding nullifies the effect of contraction stresses by having the stresses from different weld deposits working against each other. Some sequence welding techniques are as follows:

Numbers indicate sequence
of weld layers

Figure 3.16 *Balance welding*

Balance welding

In double sided multi-run fillet joints and double butt joints, welding should not be completed on one side before starting the other, but done by doing one layer on one side followed by a layer on the other (Figure 3.16).

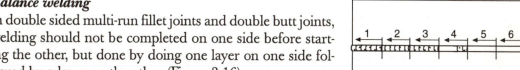

Figure 3.17 *Back step welding (numbers indicate the sequence of short passes and the arrows indicate the direction of welding)*

Back step welding

In this sequence the joint is marked out into short lengths, and after the first part is welded all other welds are done towards a welded section (see Figure 3.17). This technique is excellent for preventing transverse distortion.

Skip welding sequence

This technique is a form of balance welding and is often used to weld inserts (patches) into a plate. The joint perimeter is marked out into short lengths and welded in a sequence which allows you to skip from one part of the joint to another. If this technique is used for patching a plate, the corners of the insert should be radiused to eliminate stress points (Figure 3.18).

Figure 3.18 *Skip welding (numbers indicate the sequence of short passes and the arrows indicate the direction of welding)*

Chill blocks (artificial cooling)

When welding together parts which greatly differ in thickness, artificial cooling is sometimes used to prevent overheating the thinner part. Blocks made from a superior heat conducting material – such as copper – are placed alongside the thinner part to absorb some of the welding heat (Figure 3.19).

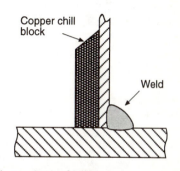

Copper chill
block

Weld

Figure 3.19 *Chill block*

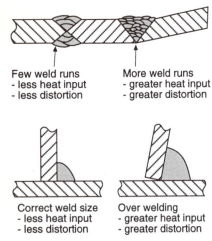

Few weld runs
- less heat input
- less distortion

More weld runs
- greater heat input
- greater distortion

Correct weld size
- less heat input
- less distortion

Over welding
- greater heat input
- greater distortion

Figure 3.20 *The number and size of weld runs*

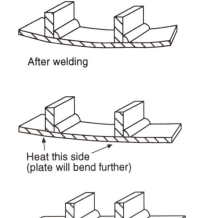

After welding

Heat this side
(plate will bend further)

Cool rapidly and
plate will straighten

Figure 3.21 *Heating and cooling*

Figure 3.22 *A dished or saucered plate*

Number and size of runs

Keep welds to the required or specified sizes. Over welding increases heat input and distortion stresses. The number of runs (passes) used to produce the desired weld size should also be kept to a minimum (Figure 3.20).

After welding

Once it has occurred, distortion is difficult to eliminate, however, some rectifying after welding is possible.

Hammering

By carefully stretching a weld face using the balled end (ballpein) of a hammer, contraction stresses can be relieved on thin material. Care must be taken not to flatten the weld face too much. A condition can be created at the toes of the weld similar to overroll, resulting in stress points which might lead to cracking in service. Hammering or peining is not generally used to rectify distortion, because of likely weld damage.

Heating and cooling

By selective heating and rapidly cooling areas of the job on the opposite side of the weld, these areas are caused to shrink thus reducing the effects of the distortion (Figure 3.21).

Distortion in flame cutting

Saucering

When shapes are flame cut from thin plate the heating and rapid cooling which occurs around the edge of the shape will cause it to distort. A good example of this condition is the dish effect (saucering) which can result when cutting discs (Figure 3.22).

Rectification

No amount of hammering or pressing in the middle of the disc or shape will rectify this condition. The dish simply springs to the other side. However, the cut edges can be stretched by pressing or hammering; the dishing will disappear, allowing the plate to flatten.

Manual metal arc welding (MMAW)

The process

In manual metal arc welding, an arc is struck between a flux covered electrode and the parts to be welded. The heat from the arc melts the surface of the metal and the consumable electrode, which crosses the arc to form part of the weld. The molten flux provides fuming and solidifies to form a protective slag over the finished weld.

Uses

MMAW is widely used in general fabrication and repair work.

It is ideal for general welding and electrodes are available for use on a wide range of steels. With practice it is an easy process to master.

Equipment

Power source

The power source for MMAW can be either alternating current (ac) or direct current (dc) Alternating current welding machines generally operate off mains power (240 V or 415 V) and are step-down transformers. These machines transform high voltage current to a comparatively low and safer level. The maximum output permissible on ac manual arc welding machines is 80 volts.

Direct current welding machines can either be run from generators or mains power using ac/dc rectifiers. The maximum output on dc machines is 115 volts. Because the voltage on some dc machines can be varied, that is, set to a particular requirement, they are often preferred to ac machines for use with electrodes which may need a higher voltage to establish (strike) the arc.

Electrode cable and holder

The electrode cable connects the electrode holder to the welding machine. It is a fully insulated and flexible copper cable. It needs to be of a large enough diameter to carry the maximum welding amperage which that machine is capable of without overheating.

The electrode holder must have an insulated handpiece and be suitable for use at the machine's maximum output.

Return cable and clamp

The return cable and the return clamp are fastened to the job, and to the welding machine. Both the cable and the clamp must be able to carry current equal to the machine's maximum output. The clamp must make good metal-to-metal contact with the job or overheating, resulting in loss of welding current, can occur.

Responsibilities of the operator

Cleaning and maintenance

The welding machine should be kept clean and dry, and cables, holders and clamps must be regularly checked for damage or deterioration. Keep cables tightly connected to the machine and make sure that the connecting lugs are not touching the machine's cabinet.

Reporting damage

Report any damage you detect to the person responsible for the machine.

Limit of repairs

Keep all cables fully insulated and holders and clamps in good condition.

Do not tamper with or repair any internal part of the machine, unless you are a licensed electrician.

Principles of operation

Voltage

Voltage is electrical pressure expressed in volts. This pushes the current through the circuit.

Open circuit voltage

Open circuit voltage is the voltage across the terminals of a welding machine,which has been set up and switched on, but measured before the arc is struck. This will be no more than 80 volts if ac equipment is used and 110 volts if dc equipment is used (Figure 4.1).

Arc voltage

Arc voltage is the voltage in the circuit measured while the arc is active. This will generally be between 20 and 23 volts (Figure 4.2).

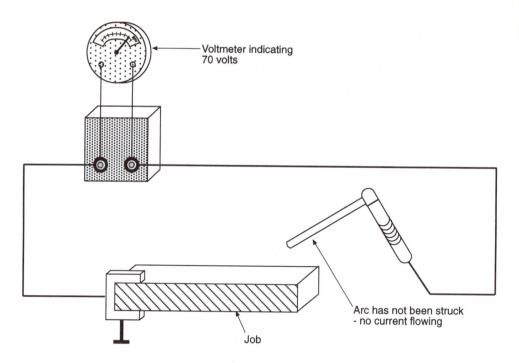

Figure 4.1 *A welding circuit being measured for open circuit voltage*

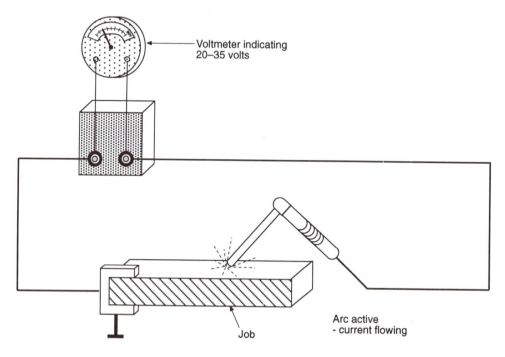

Figure 4.2 *A welding circuit being measured for arc voltage*

Amperage

Amperage is a measure of the amount of current flowing in an electrical circuit. Amperage will fluctuate with arc length; the longer the arc, the lower the amperage will be. However, because of the nature of most welding machine circuits, the percentage drop in amperage compared to the percentage increase in voltage does not significantly affect the welding performance, if the variation in arc length is due to normal hand movements during welding (Figure 4.3).

The amperage is set by the operator. For example, 130 amps is set for a 3.2 mm electrode and 170 amps is set for a 4 mm electrode.

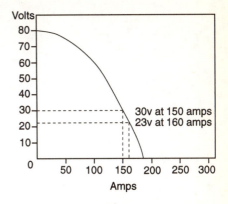

Figure 4.3 *Amperage output curve*

Metal transfer

During welding the end of the electrode core wire is transferred across the arc stream to the job. This is done irrespective of the welding position and is influenced by the following:

- Gravity has a slight and insignificant effect when using general purpose electrodes, the slag of which solidifies quickly. Electrodes with a more fluid slag are more greatly affected by gravity.
- The expansion of gases from the flux and the weld area play a big part in forcing the droplets of electrode across the arc.
- The electro magnetic forces set up by the current pinch the end off the electrode to form a droplet and propel it into the arc stream (the pinch effect).
- The surface tension of the weld pool pulls the molten droplets of electrode into the pool.

Metal transfer from the electrode to the job can only occur if the sectional area of the job is greater than the sectional areal of the electrode. If welding is attempted on smaller sectional areas than that of the electrode, transfer will be from the job to the electrode.

Functions of electrode coatings

Electrode coatings (fluxes) are made up of a variety of ingredients and are designed to improve the quality of the weld in a number of ways. Flux coatings:

- Act as deoxidisers in the weld pool preventing oxidation of the weld metal.

- Help stabilise the arc by improving its ability to carry the current.
- Produce gaseous fumes which shield the weld area from atmospheric contamination.
- Provide extra deposit metal (see iron powder electrodes later in this chapter).
- Form a protective slag which helps produce a smoother weld face and makes welding possible in positions other than flat.

Electrodes

AS1553 is the Australian standard for covered electrodes for welding carbon steels. In this standard, electrodes are classified using the letter E and a four digit number. The E signifies that it is a welding electrode and the first two digits indicate one tenth of the minimum strength of the weld deposit (Table 4.1).

For example:

E41XX = 410 MPa (minimum)

E48XX = 480 MPa (minimum)

The third and fourth digits indicate

- the type of welding current suitable
- the type of flux coating
- and the recommended welding positions.

Cellulose fluxed electrodes

These electrodes are designed for deep penetration in all welding positions. They have a fiery arc action and are ideal for root runs in butt welds on pipe.

Rutile fluxed electrodes.

Rutile electrodes are general purpose electrodes producing high quality weld, and they are easy to use and are suited to all positions.

Iron oxide fluxed electrodes

These electrodes have flux coatings which produce excellent penetration. They are especially suited to deep grooves, as the honeycomb slag they form is easy to remove.

Table 4.1 *Summary of electrodes*

Flux	Electrode classification	Suitable power	Welding position
Cellulose	EXX10	d.c., electrode positive	All
	EXX11	d.c., electrode positive or a.c.	All
Rutile	EXX12	a.c. or d.c., electrode positive or negative	All. Good for vertical down
	EXX13	a.c. or d.c., electrode positive or negative	All
	EXX14	a.c. or d.c., electrode negative or positive	All. (Contains some iron powder.)
Iron oxide	EXX20	a.c. or d.c., electrode positive or negative	Flat position and horizontal fillets only
Iron powder	EXX24	a.c. or d.c., electrode positive or negative	Flat position and horizontal fillets only
Hydrogen controlled	EXX16	a.c. or d.c., electrode positive	All
	EXX18	a.c. or d.c. electrode positive	All
	EXX28	a.c. or d.c., electrode positive	Flat position and horizontal fillets only. (Contains iron powder.)

Iron Powder fluxed electrodes

Iron powder electrodes are designed to increase welding speeds. The iron powder in the flux forms additional weld metal, giving a total weld weight of up to 130 per cent of the core wire weight. Because of their very fluid nature these electrodes are unsuitable for welds in the overhead or vertical positions.

Hydrogen controlled fluxed electrodes

Electrodes in this group produce very high quality welds. They are designed to minimise cracking when welding hard or hardenable steels, particularly medium carbon and low alloy steels. During welding, dissolved hydrogen in these steels can cause cracking in the heat affected zone of the weld (underbead cracking).

Electrode sizes

Electrodes size refers to the core wire diameter of the electrode. As the size of an electrode increases a higher amperage is required. Always refer to the electrode packet to obtain the recommended amperage range for that electrode's type, size and welding position (Figure 4.4).

Figure 4.4 *Electrode size*

Storage and handling of electrodes

- Store in a dry location off the floor.
- Use in order of receipt.
- Don't drop or handle electrodes roughly.
- Discard wet or damaged electrodes.
- Once packets have been opened, keep hydrogen controlled electrodes in an electrode oven or hot box at above 100°C.

Reconditioning

Rutile electrodes suspected of containing moisture should be reconditioned (dried) before use.

Cellulose electrodes should not be reconditioned as their flux must contain a high (7 per cent) moisture content to produce their fiery arc action.

Hydrogen controlled electrodes from previously opened packets must be reconditioned for one and a half hours at 260°C immediately before using. They should then be used directly from a hot box with its temperature kept above 100°C.

Other types of covered electrodes

Covered electrodes are available for a range of tasks in addition to welding low alloy and carbon steels.

Electrodes are available for the following operations:

- Hard facing – to produce heat , abrasion , shock or corrosion resistant surfacing.
- Welding the full range of stainless and nickel steels.
- Welding a number of non-ferrous alloys, such as aluminium and copper; however, if welding non-ferrous alloys it is best to use welding processes other than MMAW.

Application

Selection of electrode type and size

Parent metal and deposit metal must be compatible. When selecting the type of electrode, you will be influenced by the parent metal composition and the weld position. Not all electrodes can produce acceptable welds in all positions.

The electrode size must suit the thickness of the material and the access you have to the joint. A large electrode may produce oversized welds, or be unable to gain satisfactory access to the root of the joint. When welding overhead or vertically a large weld pool is more likely to collapse and be more difficult to control than a smaller one.

Amperage

Amperage will vary according to:

- electrode size
- welding position
- electrode type.

Refer to manufacturers' information and stay within their recommended ranges.

Lead and electrode angles

These are the angles at which the electrode is directed towards the joint. The transfer of metal from the electrode takes place in the direction in which the electrode is pointing. For effective welding these angles have to be closely monitored (Figures 4.5 and 4.6.).

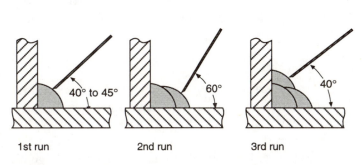

1st run   2nd run   3rd run

Figure 4.5 *Electrode angles from a 3-run, 2-layer horizontal fillet weld*

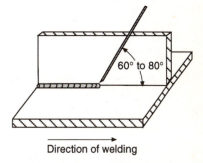

Direction of welding

Figure 4.6 *Lead angle on a horizontal fillet weld (all runs)*

Arc gap

Different electrode classifications require different arc gaps as follows:

- Contact – the touch welding technique used with iron powder electrodes where the flux coating is in contact with the metal.
- Close arc gap – where the arc gap is kept as short as possible without actually touching the job. This is used with hydrogen controlled and rutile electrodes.
- Open arc gap – a slightly longer gap is used with cellulose electrodes to maintain their fiery action and achieve deep penetration.

Arc gaps are usually not more than the electrode core wire diameter.

Preheating

When welding low alloy, medium carbon and thick or rigid low carbon steels the weld must be prevented from cooling too quickly. Moderately heating the parts to be welded, called preheating, will prevent the cooler parts of the plate from having a quenching effect on the weld. Failure to preheat can lead to the formation of brittle weld zones which are prone to cracking.

Tacking

To prevent misalignment of the parts, minimise distortion and maintain root gaps during welding, all parts should be securely tacked together before welding. Tacks should be large enough not to break under stress before welding reaches them, and in such a place that they do not interfere with the penetration or acceptance of the weld. In butt welds, tacks at the root of the preparation may need to be reduced with the edge of an angle grinding wheel before welding commences.

Deposition rate

The speed at which the joint is completed will depend on the size and type of electrode selected. Larger diameter electrodes produce larger weld deposits, and iron powder electrodes will produce larger welds than other electrode types of the same size. Care should be taken with very large diameter electrodes (greater than 4 mm) as the weld pool can easily collapse producing the weld fault overroll.

Slag removal

Slag removal between passes is essential. Failure to do this can lead to slag being trapped between runs. Welds with slag inclusions are generally unacceptable.

Problem solving

Problem	Possible cause
No arc	• power not on • return clamp not or poorly connected • bare wire portion of electrode not making contact in holder.
Electrode sticking	• amperage is too low • electrode flux damaged • voltage too low.
Arc blow *	• using dc above 200 amps.
Rough weld face	• irregular deposit rate • irregular arc length • amperage too low • incorrect electrode angles.
Excessive spatter	• arc length too long • amperage too high • old, damp or damaged electrode.
Burn through	• amps too high • incorrect preparation – root face too thin or root gap too wide.
Lack of penetration	• amperage too low • wrong electrode type/size • incorrect preparation – root face too thick or root gap too narrow.

* Arc blow is a condition which can occur when welding with d.c. current using high amperage. The arc is deflected towards one plate causing the metal transfer to take place only on that plate. This condition cannot occur using a.c. and is unusual on d.c. below 200 amps.

Problem solving (continued)

Problem	*Possible cause*
Slag inclusion	• failure to clean slag from previous run/layer • arc gap too long • amperage too low • poor joint preparation.
Undercut	• amperage too high • wrong electrode angle • welding too fast • incorrect weave technique.
Overroll	• amperage too low • welding too slow • electrode too large.
Porosity	• contaminated parent metal (rust, paint, grease) • electrode and parent metal incompatible • old or damp electrode.
Misalignment	• tacks too small or insufficient • incorrect set-up
Distortion	• unsuitable weld procedure (joint type, tacking or weld sequence).
Cracking	• weld run too small • insufficient pre-heat • wrong electrode type • joint too restrained.

CHAPTER 5

Oxyacetylene welding (OAW)

The process

In this process an oxyacetylene flame is used to melt the surface of the parts to be joined, fusing them together. The outer envelope of the flame covers the hot metal and shields it from atmospheric contamination. Additional metal may be added to the joint using a hand held filler rod. A flux is not required when fusion welding carbon steel and nickel, but is required for most other metals

Uses

Oxyacetylene welding is very versatile and almost all metals and their alloys can be welded with it. However, compared to arc welding it is slow and the fluxes needed for welding some alloys can be highly corrosive. Oxyacetylene welding is ideal for fusion welding thin steel plate and repairing parts made from cast iron. It is also widely used for brazing and braze welding iron, steel and copper alloys.

Equipment

Cylinders

Oxygen cylinders

Oxygen cylinders are made of solid drawn steel and designed to hold compressed oxygen at high pressure for industrial use.

An oxygen cylinder has a right-hand regulator connection thread and is painted black (Figure 5.1). It has a bursting disc fitted, which will blow out and let the oxygen escape if the cylinder pressure becomes too great, such as in a fire. When working with oxygen cylinders you must consider the following safety issues:

- Don't drop or handle the cylinders roughly.
- Don't use the cylinders as rollers.
- Keep cylinders cool and away from heat sources.
- Never heat cylinders with a torch or strike with an arc.
- Store and use upright on a trolley or other device to prevent cylinders from falling over.
- Don't oil gauges. If oil or grease comes in contact with compressed oxygen it can cause an explosion
- Don't use oxygen as a substitute for compressed air and never use it to dust down work areas or clothing.

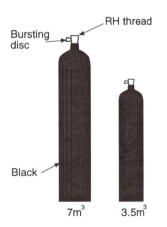

Figure 5.1 *Industrial oxygen cylinders*

Acetylene cylinders

Acetylene cylinders contain a porous material saturated with acetone in which acetylene gas is dissolved. This is similar to the way carbon dioxide is dissolved in lemonade. As a gas, acetylene is very unstable and can self detonate if it is compressed beyond 175 kPa.

Acetylene cylinders have a left-hand regulator connection thread and are painted crimson. Fusible plugs are fitted to the top of the cylinder and in the event of a fire, these melt at 100°C and allow the gas to escape (Figure 5.2).

When working with acetylene cylinders you must consider the following:

- All the precautions listed for oxygen cylinders.
- Don't pass acetylene through copper fittings or pipes. This will form copper acetylide, which is a substance capable of self detonation.
- Never set acetylene regulator pressure above 100kPa.
- To prevent drawing out acetone with the acetylene gas, restrict the draw-off rate to 1/7 of the cylinder's contents per hour.

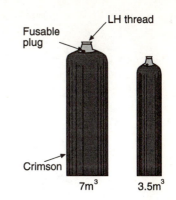

Figure 5.2 *Industrial acetylene cylinders*

Regulators

Regulators are needed to reduce and control the flow of gases from the cylinders. Oxygen regulators are coloured black and acetylene crimson (Figure 5.3).

Regulators allow:

- the contents of the cylinder to be monitored
- the reduction of cylinder pressure to a safe working pressure
- a constant working pressure to be set and adjusted.

Hoses and connectors

Reinforced rubber hoses and brass fittings are used to transport the gases from the cylinders to the handpiece handle. Hoses are black with right hand fittings for oxygen and crimson with left hand fittings for acetylene.

Handpiece

The handpiece (handle) connects to the hoses and holds the handpiece control valves (Figure 5.4).

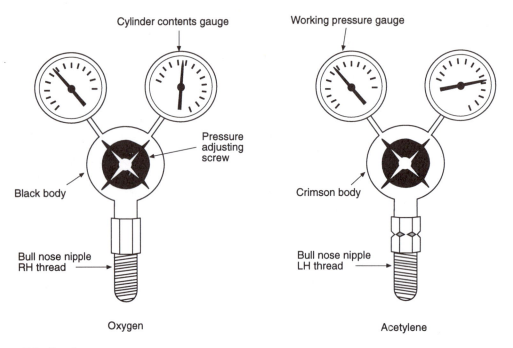

Cylinder contents gauge

Working pressure gauge

Pressure adjusting screw

Black body

Crimson body

Bull nose nipple RH thread

Bull nose nipple LH thread

Oxygen

Acetylene

Figure 5.3 *Regulators*

Mixer

The mixer screws onto the handle. It is here the gases are mixed by a spiral action as they pass through to the welding tip.

Welding tip

Welding tips direct the flame to the weld area. They come in sizes from 8 to 40 and this indicates ten times the size of the hole in the tip. For example a 12 tip has 1.2 mm hole (Figure 5.5).

Figure 5.4 *Oxyacetylene welding torch*

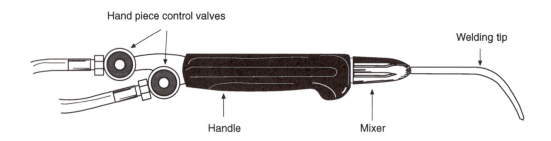

Hand piece control valves

Welding tip

Handle

Mixer

Figure 5.5 *Welding tip*

Trolley

A trolley should be used to transport the welding equipment and to prevent the cylinders from falling over.

Lighting device

An electronic gas lighter, a flint gun or a gas pilot light should be used to light the oxyacetylene flame. Matches and butane gas lighters are dangerous and must not be used.

Tip cleaners

Tip cleaners are available to keep the welding tip free of obstruction. A straight clean flame will be easier to use and produces better welds.

Principles of operation

Gas pressure limits

For welding, the oxygen and acetylene pressures need to be equal – the recommended pressures range from 50 to 100 kPa. At no time should the acetylene pressure be set above 100kPa.

Flame settings

The flame setting is critical. When welding steel, an excess of acetylene can promote hardening in the weld area, while an excess of oxygen will oxidise (burn) the metal, severely affecting its strength. There are three types of oxyacetylene flames. Each of them have uses as follows

Neutral

A neutral flame has equal amounts of oxygen and acetylene supplied to the tip. It is used for welding steel, stainless steel, nickel, cast iron, aluminium and copper. It produces a temperature of about 3100°C (Figure 5.6).

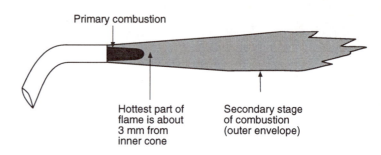

Figure 5.6 *Neutral flame*

Carburising

A carburising flame has an excess of acetylene supplied to the tip. This produces a feather beyond the inner cone. This flame is sometimes referred to as a reducing flame and is used for hard facing (Figure 5.7).

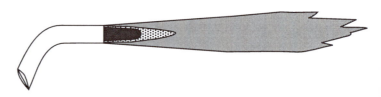

Figure 5.7 *Carburising flame*

Oxidising

An oxidising flame has an excess of oxygen supplied to the tip and burns with a hissing sound. Both the inner cone and the outer envelope are short. An oxidising flame is used for braze welding and for the fusion welding of brass (Figure 5.8).

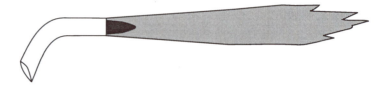

Figure 5.8 *Oxidising flame*

Additional weld metal

Additional weld metal can be added to the weld area if needed by using a compatible hand-held filler rod.

Functions of welding fluxes

Fluxes are necessary when fusion welding most metals, other than steel, to dissolve the high melting point oxides on the surface of the joint, and to prevent more from forming during the welding operation.

Forehand technique

In forehand welding the flame is always pointing towards an unwelded section of the joint. It is ideal for most welding applications, but very slow when used on steel over 4mm thick.

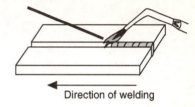

Figure 5.9 *Forehand welding*

Backhand technique

In backhand welding the flame is always pointing towards the welded portion of the joint. This prevents the weld pool from spreading on to unmelted plate and gives better penetration. Backhand welding is ideal for joints in steel more than 4mm thick because the heat concentration improves the welding speed.

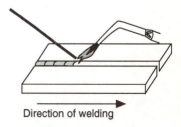

Figure 5.10 *Backhand welding*

Filler rods and fluxes

Low carbon steel filler rod

Low carbon steel filler rod (black) is a general purpose rod for welding low carbon steel, when high weld strength is not required. Joints welded with this filler rod have good ductility.

Low alloy steel filler rod

Low alloy steed filler rod (copper-coated) is a high strength filler rod suitable for joints in steel, which are subjected to stress, such as joints in pressure pipes. It produces smooth sound welds.

Cast iron filler rods

Cast iron filler rods are available for fusion welding grey cast iron. These rods have a high silicon content to compensate for any loss of silicon during welding. Without silicon, graphite flakes do not form and the casting becomes hard and brittle.

Sizes

Filler rods are supplied in 750mm lengths and diameters from 1 to 5mm.

Cast iron flux

Cast iron flux must be used when fusion welding cast iron, to clean and deoxidise the weld area. It will also prevent further contamination while welding.

Other fluxes

Copper and brass flux is needed to weld copper and copper alloys. Aluminium flux is needed when welding aluminium alloys.

Storage and handling of rods and fluxes

- Filler rods need to be covered and stored in a dry area which is off the floor.
- Filler rods should be used in the order in which they were received.
- Fluxes need to be kept in airtight containers and used in order of receipt.

Fluxes are toxic. Don't handle them with bare hands.

Application

Setting up

- Make sure the tops of the cylinders are clean. Fit the regulators, then tighten firmly.
- Fit the hoses to the regulators and handpiece, then tighten firmly.
- Fit mixer to handle and screw tip into mixer firmly.
- Loosen mixer to adjust position of tip, then retighten firmly.
- Open the cylinder valve wheels one and a half turns to allow gases into the regulators.
- Open the oxygen handpiece valve and turn the regulator adjusting screw until the desired pressure registers on the gauge.
- Close handpiece valve. If a slight creep registers in the pressure reading, do not adjust; this will drop again when in use.
- Repeat this procedure with the acetylene.
- To test for leaks, brush fittings with soapy water while the system is pressurised. Bubbles will indicate leakage. Retighten or replace fittings if necessary.

Selection of tip size

The size of the tip selected will depend on the thickness of the metal being welded. Welding will be too slow if too small a tip is selected and overheating will be a problem if too large a tip is used.

Selection of filler rod type and size

Filler metal must be compatible with the parent metal. Other factors which will influence the selection of the type of filler rod are:

- strength requirements
- corrosion resistance
- colour matching properties.

The size of the filler rod selected will be influenced by the metal thickness and generally, the filler rod will be equal to or less than the thickness of the metal being welded.

Edge preparation

Plates up to 3mm thick will not need any special edge preparation, however, very thin metal will be easier to weld if a small lip is put on the parts forming the joint. Plate over 3mm thick may have to be bevelled. Bevelled joints for oxyacetylene welding are wider, have no root face and a root gap is not usually needed. Plates up to 8mm thick can be welded using open square butt joints if the backhand technique is used (Figures 5.11 and 5.12).

Gas pressures

For tip sizes 8 to 26, 50 kPa for both oxygen and acetylene is recommended and for tips above this, 100 kPa for each gas is required.

Flame adjustment

After lighting the acetylene its flow is adjusted so that the flame is in contact with the tip, it has a strong flow and most of the thick black smoke has stopped. The oxygen is then slowly added, until the flame changes from red to blue and then to a bright blue cone which is surrounded by a whitish feather. This feather will become shorter as the oxygen flow is increased. Continue to add oxygen until the feather just meets the inner cone. This is a neutral flame (Figure 5.13).

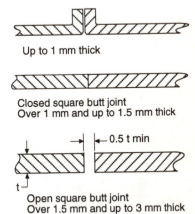

Up to 1 mm thick

Closed square butt joint
Over 1 mm and up to 1.5 mm thick

0.5 t min

Open square butt joint
Over 1.5 mm and up to 3 mm thick

Figure 5.11 *Joints for metal up to 3 mm thick*

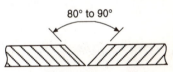

80° to 90°

Figure 5.12 *Single V joint*

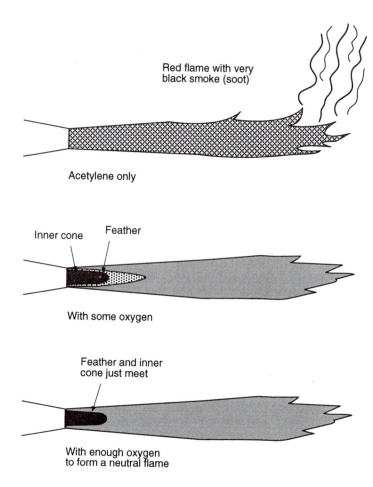

Red flame with very black smoke (soot)

Acetylene only

Inner cone   Feather

With some oxygen

Feather and inner cone just meet

With enough oxygen to form a neutral flame

Figure 5.13 *Neutral flame adjustment*

Practise setting this flame and make sure that you do not add too much oxygen or you will get an oxidising flame. A poorly set flame will make the welding difficult and reduce the quality of the weld.

Flame-to-job distance

To increase welding speed, 3mm is the best flame-to-job distance. The hottest zone in a neutral flame is about 3mm in front of the cone.

Weld pool

The weld pool should be clear and tranquil and appear wet. If boiling or burning are evident check your flame settings. If scale and dirt appear to be floating on the pool the metal or filler rod are not clean.

Filler rod manipulation

When welding, the filler rod should not be removed from the protection of the outer envelope of the flame, except to add more flux to it on those jobs which require flux. The filler rod is repeatedly dipped into the weld pool to melt it as the job progresses. On fillet welds a slight brush on to the vertical plate can help prevent undercut. A stirring action (puddling) is best used on cast iron to release gas from the pool. Cast iron jobs must be preheated to a dull red, heated again after welding and allowed to cool as evenly and slowly as possible to prevent them becoming brittle.

Welding tip and filler rod angles

See Figure 5.14

Weld craters and re-starts

When the progress on a weld is broken (halted) a flat crater forms. This crater is very thin and prone to cracking, so before finishing a weld or when restarting one, additional weld metal should be deposited to build the crater up to the full weld thickness.

Control of penetration

The volume of heat from the welding tip must be sufficient to melt the root of the joint. Penetration through the root is fairly easy to control on carbon steel and cast iron, because the surface tension of these metals when molten is comparatively strong and holds the penetration bead in place. On non-ferrous metals, such as aluminium, there is no colour change when they are heated and a sudden collapse of the penetration bead is common. However, copper support plates can be used to shape and hold penetration beads when welding aluminium (Figure 5.15).

When welding copper plate full penetration can be difficult, as copper exhibits sudden melting and rapid solidification. A tip size one larger than would be used on steel of the same thickness is required.

Appearance of ripples

The uniformity of the ripples on the face of an oxyacetylene weld indicates the steadiness of the filler rod manipulation and the progress along the job. Rough weld surfaces can lead to failure under service stress.

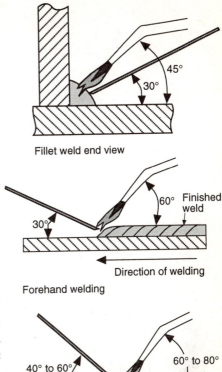

Fillet weld end view

Forehand welding

Backhand welding

Figure 5.14 *Tip and rod angles*

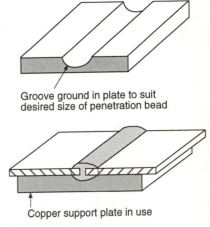

Groove ground in plate to suit desired size of penetration bead

Copper support plate in use

Figure 5.15 *Penetration support*

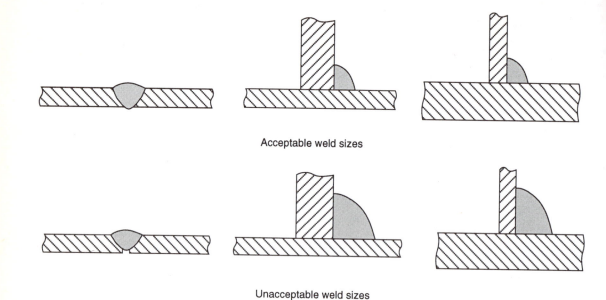

Acceptable weld sizes

Unacceptable weld sizes

Figure 5.16 *Weld sizes*

Weld size

Weld size is governed by plate thickness, and the throat thickness of butt joints should be not less than plate thickness. The prepared joint gap must be completely filled. The leg length of fillet welds should equal the plate thickness and when plates of different thicknesses are fillet welded, the weld size must suit the thinner material (Figure 5.16).

Tacking

All jobs should be fully tacked before welding commences. Failure to do this may result in distortion which will make further assembly difficult.

Tacks should be large and frequent enough to prevent them breaking or allowing misalignment of the assembled parts.

Special conditions

Flash back

Flash back is a situation where the gases have become ignited inside the welding handle; the welding tip appears to be extinguished and there is a shrill hissing sound. This condition is potentially dangerous. Should it occur the

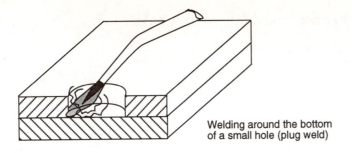

Welding around the bottom
of a small hole (plug weld)

Figure 5.17 *A sectioned view of a situation which can overheat the welding tip creating a flashback.*

oxygen must be turned off immediately, then the acetylene. The torch should then be checked for damage to the 'O' rings in the mixer and control valves.

Anti-flash back fittings are available. These have a floating disc which closes off gas supplies to the tip in the event of a back-pressure. They are fitted between the handpiece and the supply hoses.

A flash back is caused by the welding tip overheating and could happen if welding is attempted where the heat is deflected back into the tip.

Backfire

A backfire occurs when the flame is momentarily extinguished and then relit from the hot plate. This is accompanied by a loud bang and a shower of sparks. It is not as dangerous as a flash back but is quite unnerving for the beginner.

After a backfire has been experienced the flame should be turned off, the tip cooled and then cleaned before further welding is done. The causes of backfiring are:

- A dirty tip.
- Insufficient gas flow (too soft a flame).
- An overheated tip
- The tip touching the job.
- Scale or other surface contaminant lifting and blocking the tip.

Problem solving with OAW

Problem	*Possible cause*
Weld pool too slow to form	• tip too small • flame setting too soft.
Filler rod sticking	• weld pool too cold • filler rod not being directed into weld pool.
Excessive sparking	• oxidising flame instead of a neutral flame • flame too harsh • dirty plate surface or filler rod.
Burn through	• flame too harsh • tip too large • insufficient filler rod being applied • root gap too wide.
Lack of penetration	• tip too small • flame too soft • root gap too small • welding too fast • filler rod too large.
Undercut	• unsatisfactory rod manipulation or insufficient filler rod • tip too large • flame too harsh • excessive weaving of welding tip • wrong tip and filler rod angles • welding too fast.
Overroll	• insufficient heat • weld pool too large
Porosity	• wrong filler rod • unclean plate or filler rod • not enough stirring action in weld pool (cast iron)

Problem solving with OAW (continued)

Problem	*Possible cause*
Lack of fusion	• tip too small • flame too soft • filler rod too large • poor joint preparation.
Misalignment	• incorrect set-up • tacks too small.
Cracking	• wrong flame setting • welds too small • job too rigid • wrong filler rod type • job cooling too quickly.
Distortion	• tacking too small or infrequent • wrong weld sequence.

Brazing

The process

Brazing uses an oxyacetylene flame to heat pre-fluxed and close-fitting parts which are to be joined, and to melt the filler rod. The parent metal is not melted and the filler metal is drawn between the close-fitting parts by capillary attraction. Capillary attraction is the term used to describe the way liquids will flow between certain surfaces and into small crevices, even though this may be against the pull of gravity.

When a silver alloy is used, the process is called silver soldering or silver brazing.

Uses

Brazing is ideal on thin metal because the lower heat used causes less distortion. Before deciding on the suitability of this process you must consider the following:

- Strength – it is not a fusion weld.
- Cost – some filler rods are very expensive.
- Type of joint – brazing is strongest when there is maximum contact between the parts. Normal fillet or butt joints are not generally suitable joints for brazing.

Brazing is commonly used to join ferrous and non-ferrous metals (especially copper) in the refrigeration and electrical industries.

Joints

See Figure 5.18.

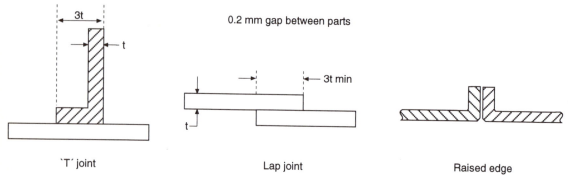

Figure 5.18 *Some joints suitable for brazing*

Cleaning of parts to be joined

Thorough cleaning is essential before brazing. Parts should be buffed with steel wool and degreased using acetone or white spirits.

Filler rods

A wide range of brazing rods are available. Some of these are listed below:

Type of brazing rod	*Use*
Aluminium brazing rod	For brazing pure aluminium
Phosphorus copper brazing alloy	For the fluxless brazing of copper
Silver brazing alloys (silver content ranging from 2 to 55%)	For brazing most ferrous metals, copper and copper alloys (brass and bronze)

Flux

The flux must suit the filler rod and the parent metal. Accordingly, a general purpose brazing flux is available to use with silver brazing alloys, and a special aluminium flux must be used when brazing aluminium.

Technique

Flux application

Flux is applied in paste form to the weld area only. If flux is applied to other parts of the job, filler metal will also flow onto these areas, where it is not needed. Flux is also applied to the end of the brazing alloy rod.

Flame setting

A soft neutral flame is required. See Figure 5.6

Heating

The flame is not held closely to the metal to be joined as it is with fusion welding. The parts should be heated with the flame moving along the joint continuously with a 'washing' action.

As the metal nears its brazing temperature, the flux will begin to dry out and become powdery and white. On further heating the flux will melt, becoming transparent and take on a wet look.

Application of filler rod

When the flux has melted the metal in the joint itself should be hot enough to melt the filler rod. With the flame still moving, the joint is touched with the filler rod, melting a small portion of the rod. This is worked along the joint with the flame. More of the filler rod is added until the joint is complete. Metal from the filler rod should now be visible on the other side of the joint.

Flux removal

Flux residue is corrosive and should be removed by using hot water and wire brushing.

Braze welding

The process

Braze welding uses an oxyacetylene flame to heat the parts to be jointed and to melt the filler rod. The parts to be welded are not melted but unlike brazing, the joints and application are very similar to fusion welding. The bond is achieved by intergranular penetration. This is when the filler metal flows between the grain structure of the heated surfaces.

Uses

Braze welding produces very strong welds on cast irons, carbon steels, copper and copper alloys. It is ideal where heat input is best kept to a minimum, such as joining cast irons. It does however have some limitations. Braze welding is unsuitable where:

- Service temperatures of the job are above 260°C.
- The job is exposed to ammonia.
- The joint is required to be under compression.

Joints

Joints suited to braze welding are the same as those suited to fusion welding; fillet welds and V joints require the same preparation.

Braze welding filler metal is very fluid when molten, so for this reason joints on metal more than 3 mm thick are best done in the semi-vertical position.

Filler rods

Three types of braze welding filler rods are available and although they are actually brass they are called bronzes:

Filler rod	Use
Tobin bronze	braze welding steel and copper, and for fusion welding brass.
Manganese bronze	braze welding grey cast iron.
Nickel bronze	higher strength brazing of steel and for providing a good wearing surface in bushes.

Fluxes

Copper and brass flux is used when braze welding carbon steel, copper and copper alloys. Cast iron braze welding flux must be used on cast irons.

Technique

Edge preparation

All edges to be braze welded must be clean.

The joint geometry is the same as for fusion welding, however, edges on castings should be chipped and filed rather than ground. Grinding smears graphite across the joint faces reducing intergranular penetration and bonding.

Flux application

The end of the filler rod is heated and dipped into the powdered flux. This causes some of the flux to adhere to the hot filler rod. The flux is rarely applied directly to the joint and pre-fluxed filler rods are available.

Flame setting

Slightly oxidising. See Figure 5.8.

Heating

The parts to be joined need to be heated to above the melting point of the filler rod, which will be somewhere between 800 and 900°C. Steel and iron will have become a dull red by this time.

On metal welded in the semi-vertical position the torch will be directed slightly uphill and follow the filler rod rod along the joint.

Application of filler rod

The filler rod is melted by the flame from the welding tip as the job is heated, and it is deployed in a similar way as it is applied to the fusion welding of steel. It is only removed from the protection of the flame when more flux is needed.

On semi-vertical joints a triangular weave is used to complete the joint in a single pass. The welding tip follows the filler rod during this weave.

Flux removal

Removal of the glass-like flux residue is essential. This can be done by using hot water and wire brushing.

CHAPTER 6

Flame cutting

Chapter contents

The process

When carbon steel is heated and exposed to a pure oxygen environment it will oxidise (burn) very rapidly. The oxidising of steel takes place at a much lower temperature than its melting point. Special flame cutting equipment allows steel to be oxidised and the molten oxide removed. A narrow slot, known as a kerf, is cut through the metal. With practice, flame cut edges can be clean and accurate.

Equipment

Oxyacetylene or oxy LPG plant

Oxyacetylene or oxy LPG cutting equipment uses the same cylinders, regulators and hoses as those used for heating and welding. However, because of their different pressure/flow ratios, acetylene and LPG regulators are not interchangeable.

Cutting torch

A cutting torch, fitted to the hoses in place of a welding torch, allows the setting and control of the preheating flame and the cutting oxygen, needed to oxidise the steel. The cutting oxygen is controlled by a thumb operated lever on the torch handpiece (Figure 6.1).

A cutting attachment can be fitted onto the end of the welding handpiece after the mixer has been removed and operates in the same way as a cutting torch, but has an extra oxygen control valve (Figure 6.2).

Cutting nozzles

Nozzles designed for cutting have a group of preheating oxygen holes in a circle around a larger cutting hole. The size given to a nozzle is the diameter of the cutting oxygen hole. LPG nozzles have a recessed face to make lighting and maintaining the flame easier. Acetylene nozzles cannot be used with LPG (Figure 6.3).

Tip cleaners

Stainless steel reamers are available for cleaning cutting nozzles. For clean accurate cutting the cutting oxygen hole needs to be regularly cleaned. Any deflection of the cutting oxygen stream will result in the steel being oxidised in that direction, and the quality of the cut may not be acceptable.

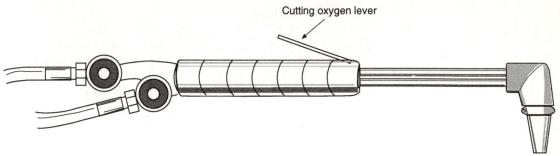

Cutting oxygen lever

Figure 6.1 *Cutting torch*

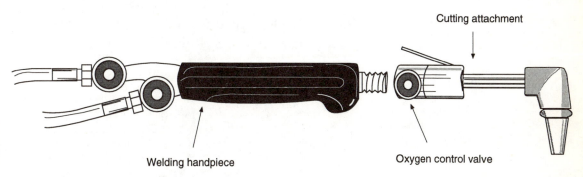

Cutting attachment

Welding handpiece

Oxygen control valve

Figure 6.2 *Cutting attachment (screws onto handpiece)*

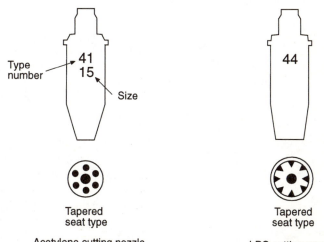

Type number

41
15

Size

Tapered
seat type

Acetylene cutting nozzle

44

Tapered
seat type

LPG cutting nozzle

Figure 6.3 *Acetylene cutting nozzles*

Cutting guides

Guide wheels

Completely freehand cutting takes a lot of practice to achieve a high degree of quality and accuracy. The grooves which can be seen on the edge of a cut are the result of the operator's hand movement. Such movement can be reduced by fitting guide wheels to the nozzle or by using other cutting aids (Figure 6.4).

Angle bar guide

This can be used to produce cuts at 90° through the plate and to cut a 45° bevel.(Figure 6.5).

Circle cutting attachment

By removing one of the wheels on the guide wheel attachment and replacing it with a longer bar, a guide suitable for accurate circle cutting is obtained (Figure 6.6).

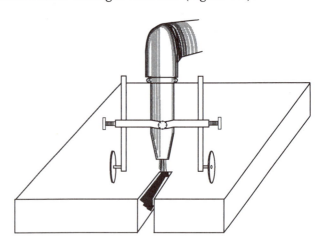

Figure 6.4 *Guide wheels*

The nut at the nozzle
is slid along the angle

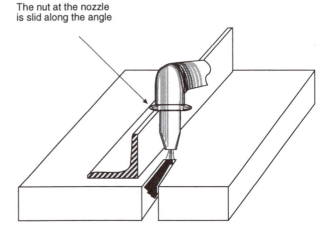

Figure 6.5 *Angle bar guide*

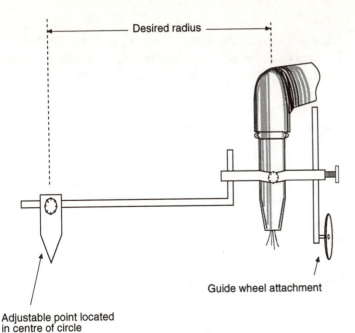

Desired radius

Guide wheel attachment

Adjustable point located
in centre of circle

Figure 6.6 *Circle cutting attachment*

Principles of operation

Ignition temperature

This is the temperature at which something will burn (oxidise) if oxygen is available. The ignition temperature for steel is 815°C. Steel must be at this temperature before the cutting oxygen can oxidise it. Carbon steel begins to turn a dull red at about 820°C so judging the ignition temperature at the initial preheat phase is relatively easy. Once a small section on the surface of a plate has been preheated to ignition temperature and oxidised, heat from the oxidising process rapidly heats the metal adjacent to it in the path of the oxygen stream. This is also oxidised and so on until the entire thickness of metal is severed. The oxidised metal (oxide) is melted by the flame and then removed by the force of the cutting oxygen stream.

Flame setting

The preheat flame should be neutral. An oxidising flame will deteriorate the quality of the cut, while a carburising flame may not produce sufficient preheating.

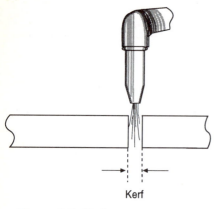

Figure 6.7 *Kerf*

Kerf

This is the space created when the cut is made. The vertical grooves on the edges of the kerf are a good indication of the suitability of the gas pressure (Figure 6.7).

Limitations

Flame cutting using conventional cutting equipment is restricted to carbon and carbon/manganese steels.

For a metal to be readily cut it needs to have:
* an ignition temperature below its melting point
* an oxide melting point below its melting point.

This is characteristic of iron and only some iron based metals can be cut.

Flame cutting safety

During cutting large volumes of molten oxide are created. You must protect your body, face, hands, feet and eyes (wear appropriate cutting goggles). Burns from flame cutting can be very severe.

Application

Selection of nozzle type and size

The type of nozzle must suit the fuel gas being used (LPG and acetylene nozzles are not interchangeable). Cutting nozzles must be large enough to supply sufficient gases for preheating and oxidising. The thicker the section being cut, the larger the cutting nozzle needs to be (Table 6.1).

Regulator pressures

During cutting using either LPG or acetylene, the same regulator pressures are used. Operating pressure restrictions are the same when using acetylene for welding or cutting.

The thicker the section being cut the more oxygen you will need to oxidise it, so the oxygen regulator pressure will have to be set progressively higher. Using a larger tip, at the same pressure, will increase the oxygen supply.

Flame adjustment

Open the acetylene control valve on the handpiece and light so that a steady flame is produced.

Table 6.1 *Tip sizes and gas pressures for given plate thicknesses*

Oxyacetylene Cutting				
Plate thickness (mm)	Nozzle sze	Oxygen pressure (kPa)	Acetylene pressure (kPa)	Cutting speed (mm/sec)
6	8	180	100	6
12	12	200	100	5
20	12	240	100	5
25	15	180	100	4
50	15	300	100	3
100	20	350	100	2
Oxy LPG cutting				
6	8	180	100	6
12	12	200	100	5
20	12	240	100	5
25	15	300	100	4
50	15	350	100	3
100	20	350	100	2

Slowly open the oxygen control valve until a neutral flame is formed.

Hold down the cutting oxygen lever and re-adjust the flame to neutral, if necessary.

Release cutting oxygen lever.

The flame is now ready to use.

Cutting attachments have two oxygen control valves. The one on the welding handpiece should be opened fully before lighting the acetylene. The one on the cutting attachment is used to make flame adjustments.

To close down the torch, turn off the acetylene valve then the oxygen valve, close the cylinder valves and release pressure on regulator knobs and hoses. Close all valves.

Nozzle to job distance

To prevent nozzle damage and overheating of the top edges of the kerf during cutting, the nozzle should be kept about 8 mm from the plate (Figure 6.8).

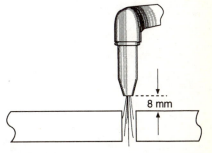

Figure 6.8 *Nozzle to job distance*

Preheating and starting the cut

To make a cut from the edge of a plate, the surface of it at that point is heated to a bright red. The cutting oxygen lever can now be depressed and the cutting started.

On very thick plates or when the weather is cold, the plate can be warmed up by passing the torch along the cutting line before commencing the procedure outlined above. Warming these jobs before cutting will improve the quality of the cut and ensure that the cutting process does not stop due to heat loss.

Cutting speed

The cutting speed should be maintained to produce the best quality cut. Attempting to cut too quickly for the nozzle size and plate thickness will result in a rough and irregular cut. Cutting too slowly can cause the top of the plate to overheat and the edges will be rounded or otherwise damaged. For approximate cutting speeds, refer to Table 6.1.

Piercing

Sometimes it is necessary to pierce the plate some distance from its edge. If this is not done carefully, permanent damage can be done to the nozzle by the oxide, which is propelled upwards at the nozzle before the plate is pierced.

The procedure for piercing to form a hole is:

- Hold the torch near the centre of where the hole is required at the usual cutting height of 8 mm until that spot on the plate is bright red.
- Raise the nozzle 25 mm from the job and slowly open the cutting oxygen lever.
- Slowly lower the nozzle to its normal height and move it along the desired cut line as the plate is pierced.

An acceptable cut

An acceptable cut will have:

- square top edges
- straight kerf
- smooth surfaced edges free of deep grooves
- no adhering slag. (Figure 6.9).

Smooth daglines

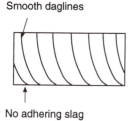

No adhering slag

Square edges

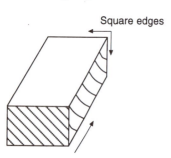

Straight surface

Figure 6.9 *An acceptable cut*

Problem solving

Problem	Possible cause
Rounded top edges	• too much preheat • nozzle too large • cutting too slow • holding nozzle too close to the plate.
Parts won't separate	• cutting too fast • nozzle too small • dirty nozzle.
Adhering slag	• dirty nozzle • cutting too slow.
Sloping kerf	• holding the nozzle at the wrong angle.
Grooved cut surface	• cutting oxygen pressure too high • too much preheating • dirty nozzle.
Irregular drag lines	• cutting speed too fast.

Flame gouging

The process

Flame gouging is an adaptation of the flame cutting process. Flame gouging nozzles provide a large volume of oxygen to the preheated surface, but at a low velocity. This means that a wider area of plate is oxidised and a clean smooth groove is cut into the surface of the plate without severing it. The depth and width of the groove are controlled by the angle of the nozzle and the number of passes made across the plate.

Uses

Flame gouging can be used to remove excess metal from a plate surface in order to:

• prepare plate edges for U joints
• remove faulty welds
• remove the penetration bead on butt welds so that a backing run can be done. Figure 6.10.

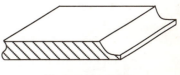

Flame gouged plate for a U joint

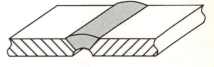

Penetration bead removed from a butt weld

Figure 6.10 *A flame gouged joint*

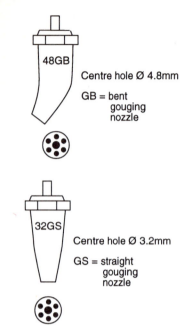

Figure 6.11 *Gouging nozzles*

Equipment

Any oxyacetylene or oxy LPG cutting plant. All the equipment used for cutting – other than the nozzle – is the same as that used for flame gouging.

Nozzles

Flame gouging nozzles have a very large centre hole. There are straight and bent types and they are fitted to the cutting head of a normal cutting torch (Figure 6.11).

Application

Nozzle selection

The width and depth of metal which you have to remove will determine the size nozzle you need. Nozzles are available with centre orifices of 3.2, 4.8 and 6.4 mm.

Gas pressures

Compared to cutting, very high oxygen pressures are needed for flame gouging, however fuel gas pressures are lower (see Table 6.2).

Flame adjustment

A neutral flame is required and the procedure for setting is the same as for cutting. However, a strong flame should be set to prevent it going out when the oxygen lever is depressed or released.

Starting the gouge

To start a gouge the plate surface or edge must be preheated to its ignition temperature as with normal cutting. After preheating the oxygen lever is slowly depressed and the gouging commenced.

Nozzle Angle

During gouging the nozzle is held at a low angle (7–15°) to the plate surface. If a deeper gouge is required this angle can be increased directing the oxygen deeper into the plate. During preheating when gouging is not commencing from an edge, a steeper nozzle to plate angle is used to prevent the preheating flame from being deflected off the surface (Figures 6.12 and 6.13).

Table 6.2 *Regulator pressures in kPa*

Nozzle size	Oxygen	Acetylene	Width of groove (mm)
32GS	500 to 550	70	8
48GB	550 to 620	70	10
64GB	600 to 700	70	13
	Oxygen	**LPG**	
32GB	500 to 550	100	8
48GB	550 to 620	100	10
64GB	600 to 700	100	13

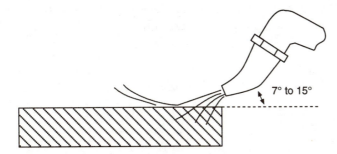

7° to 15°

From an edge

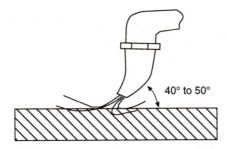

40° to 50°

On a flat surface

Figure 6.12 *Angle of nozzle during preheating*

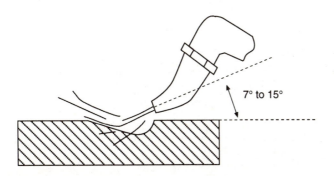

7° to 15°

Figure 6.13 *Angle of nozzle during cutting*

Caution: Take care where you direct your gouging. Much of the molten oxide is moving horizontally and not downwards at the floor as in flame cutting, and can easily start fires and damage glass and finished surfaces.

Problem solving

Problem	Possible cause
Backfiring	• flame too soft • insufficient oxygen pressure.
Irregular groove	• nozzle to plate angle not constant • dirty plate • dirty nozzle.
Gouge too deep	• nozzle to plate angle too steep • nozzle too large.
Gouge too shallow	• nozzle to plate angle too small • nozzle too small.
Too much adhering slag	• nozzle not pointed squarely in the direction of travel • dirty plate • dirty nozzle.

CHAPTER 7

Gas metal arc welding (GMAW)

The process

In this process an arc is established between a continuous bare wire electrode and the job. The heat of the arc melts the surface of the job and the electrode, which is constantly propelled into the weld pool, provides additional weld metal. Atmospheric contamination of the weld area is prevented by a shield of gas; this gas is supplied simultaneously with the electrode wire.

A lot of the work previously done with the manual metal arc welding (MMAW) process is now being done with GMAW. With better technology and lower equipment costs, some of the drawbacks associated with the process have been overcome. The introduction of small units operating on 240 volt mains power have added to the popularity of this process for non-industrial applications. More user-friendly machines have made GMAW much easier to understand and use.

Uses

GMAW can be used to weld almost all industrial metals as long as a wire compatible with the parent metal is used. The process is ideal for welding sheet steel because it allows good control and a lower heat input than that used in MMAW, reducing distortion.

Equipment

Power source

A direct current with the electrode positive (dcep) must be used as a power source for all GMAW applications. Alternating current can be changed to satisfy GMAW requirements by using an ac/dc rectifier type welding machine.

Return (earth) cable and clamp

As with all arc welding processes the welding circuit cannot be completed unless the return cable connects the machine to the job.

Wire feed unit

The wire feed unit houses the drive rolls and the spool of electrode wire. The drive rolls are grooved to suit the type of wire and they must be adjusted to prevent them slipping or flattening the wire. This continuous electrode is propelled via the conduit to the handpiece, which is called a **GMA** welding gun.

Shielding gas system

The system which supplies the shielding gas comprises a cylinder of appropriate gas, a gas heater, a regulator and a flow meter. After passing through these the gas is then fed into the conduit connected to the gun.

Conduit

The conduit delivers the shielding gas and electrode wire to the welding gun. It has a liner so that the electrode wire moves through it smoothly. At no time should the conduit be kinked or bent sharply, because this can stop or slow down the electrode feed, and may cause permanent damage to the liner.

Welding gun

The welding gun controls the power to the electrode and the drive rolls, and the gas flow. When the trigger is squeezed the contact tip at the nozzle of the gun allows power to pass to the electrode, starts the drive rolls motor to propel the wire, and opens a solenoid which allows the shielding gas to flow. When the trigger is released all activity ceases. The nozzle surrounding the contact tip must be kept clean of spatter, which can interfere with the flow of the shielding gas, and can cause a short circuit between the contact tip and the nozzle.

Figure 7.1 *A 240 volt Transmig 250 Compact, ideal for rural areas where 415 volt power is not available (courtesy of CIGWELD)*

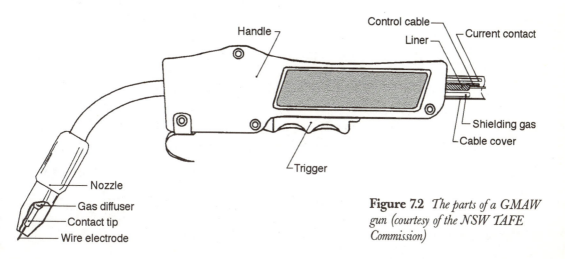

Figure 7.2 *The parts of a GMAW gun (courtesy of the NSW TAFE Commission)*

Principles of operation

Voltage

The voltage must be adjustable to perform gas metal arc welding and will vary from 16 to 40 volts according to the transfer mode required (see below).

Amperage and wire speed

Amperage (current) and wire speed are linked and any increase in wire speed will automatically increase the amperage output.

Metal transfer modes

Changes to the voltage and amperage will alter the manner in which the molten electrode crosses the arc (metal transfer). The type of transfer you need must suit the wire diameter, the plate thickness and the position of the weld.

Short arc transfer

For short arc transfer the electrode wire is pushed from the gun to the surface of the job. A short circuit occurs, overheating the thin wire which melts and is drawn into the weld pool. The arc is re-established, but the wire soon reaches the weld pool again and short circuits once more. This occurs some 200 times a second and is responsible for the familiar crackling sound associated with short arc transfer. This mode is suitable to use in all positions. Short arc transfer occurs when using 16-23 volts below 200 amps (Figure 7.3).

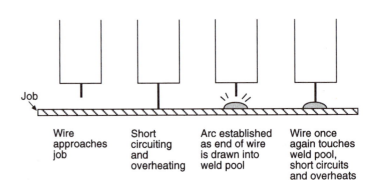

Figure 7.3 *Short arc transfer*

Spray transfer

Spray transfer increases the deposition rate. It can only be used on most metals in the flat and horizontal fillet positions. In this mode a stream of small droplets is propelled along the column of the arc from the electrode wire to the job. Spray transfer occurs with settings of 26 to 40 volts with above 200 amps. It is used to weld plates over 6 mm thick (Figure 7.4).

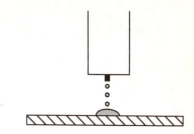

Figure 7.4 *Spray transfer*

Globular transfer

Globular transfer is a third way in which the filler wire can be transferred across the arc. It occurs at voltages between short arc and spray transfers. That is, at more than 23 volts and less than 26 volts. It is difficult to avoid these settings when welding plates around 5 mm thick. The droplets it produces are not round and the weld can have a rougher appearance than those done with short arc or spray transfers (Figure 7.5).

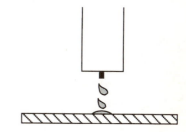

Figure 7.5 *Globular transfer*

Welding positions

All positions can be employed when using short arc transfer, but only flat and horizontal fillets are possible with spray transfer.

Shielding

Shielding is needed to prevent contamination of the weld by the atmosphere. When working outdoors care must be taken to protect the gas shield from the wind.

Electrode wires

Electrode wires can be solid or tubular (flux-cored).

Solid wires

The diameters of the most popular solid wires used on steel are 0.8, 0.9 and 1.2mm. Other sizes are available but are not as common. Wires are also available for stainless steel, aluminium, copper, nickel and alloys. Solid steel wires are copper coated to prevent rust and to provide better electrical pick-up from the contact tip.

Flux-cored wires

Flux-cored wires are like a MMAW electrode inside out and the folded tube is packed with flux. When used, a slag is formed by the molten flux, which has to be removed in the same way as with manual metal arc welding. The advantage of these wires is their very high deposition rate. The flux is the iron powder type which provides additional weld metal, however, they can only be used in the flat or horizontal fillet position. Some flux-cored wires are self shielding and require no additional shielding gas.

Storage and handling

Store electrode wires in unopened packets in a dry place off the ground and use in order of receipt. Full rolls of wire are heavy and care should be taken when reloading wire feeders.

Shielding gases

Different metals require different shielding gas mixtures. There is a very big range of gas mixtures available and your supplier should be contacted if you are in doubt. Table 7.1 is provided as a guide.

Table 7.1 *Shielding gases*

Shielding gases	Comments
Carbon dioxide (CO_2)	A low cost gas producing a very hot weld pool. It is used on low carbon steel, but gives high spatter levels.
Argon + CO_2	Lower spatter levels on steel. The CO_2 reduces cost while maintaining heat input.
Argon + oxygen	Used on steel to provide better weld properties. Used in spray transfer of stainless steel.
Argon + helium + hydrogen + CO_2	A special gas for short arc transfer on stainless steel.
Argon	Used to weld aluminium and its alloys. It is also suitable for copper and nickel alloys.
Argon + helium	Increases heat input. Used on thick aluminium and copper.

Application

Protective equipment and clothing

As with all arc welding, your body, hands, feet and head must all be protected from heat and radiation. Your head shield must have an appropriate lens fitted. Work in a well ventilated area.

Before you start

Check:

- That your electrode wire is compatible.
- That your shielding gas is suitable and that the flow meter has been set.
- Voltage, amperage and wire speed. Make sure that these will produce the transfer mode you require.

Direction of Welding

It is usual to have the gun is pointing in the direction of travel (pushing) to ensure better shielding. Welding can be done with the gun pointing away from the direction of travel (dragging), but problems with shielding may arise (Figure 7.6).

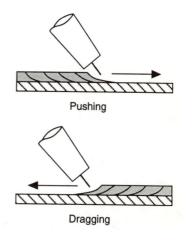

Pushing

Dragging

Figure 7.6 *Weld directions*

Gun angles

Figure 7.7 is a guide only, however, electrode angles are similar to those used in MMAW.

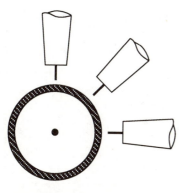

Butt weld on pipe in the fixed position. The gun should always point to the centre of the pipe

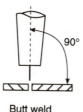

90°

Butt weld

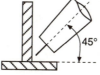

45°

Fillet weld

Figure 7.7 *Gun angles*

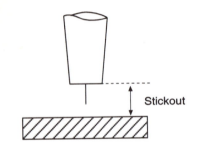

Figure 7.8 *Stickout*

Stickout

This is the distance from the contact tip to the job. If the stickout is too long the welding will become erratic and spatter levels will increase, and there is also a greater chance of poor shielding. Variations in stickout due to normal hand movements are compensated for by the machine keeping the arc gap almost constant (Figure 7.8).

Tacking

To prevent misalignment and distortion all parts should be tacked together before welding. Tacks have to be large enough and suitably spaced for the thickness of metal being welded.

Craters and restarts

All craters must be filled to the full size of the required weld. Restarts should be smooth and without excessive variation in the throat thickness of the weld.

Run-on tabs

To eliminate cold starts on weld runs, run-on tabs can be tacked to the metal being welded at the beginning of a joint. Welding is started there and progresses uninterrupted onto the job. They can be removed by flame cutting after the job is completed (Figure 7.9).

Figure 7.9 *Run-on tabs*

Preheating

Preheating may be necessary when the short arc mode is used on thick or high heat conducting metals. The unacceptable rapid cooling which can occur without preheating can lead to weld failure.

Problem solving

Problem	Possible cause
No arc	• power not on • power box fuse blown or circuit-breaker open • control fuse blown • fault in gun trigger • poor return clamp contact.
Erratic wire feed	• damaged liner • sharp bend or kink in conduit • dirty or damaged contact tip • poorly adjusted or worn drive rolls.
Uneven weld appearance	• too high a gas flow • stickout too long • wrong shielding gas • current (wire speed) too high.
Excessive spatter	• voltage too high • damaged or blocked nozzle • wrong shielding gas.
Burn through	• wire speed (current) too high.
Lack of penetration	• wire speed too low • stickout too long • root gap too small • dirty wire of contact tip.
Undercut	• wrong gun angle • welding too slow for wire speed • voltage too high.
Porosity	• dirty plate • contaminated gas • blocked nozzle • gas flow too great • incompatible wire.
Cracking	• insufficient preheat • welds too small • job too rigid.
Distortion	• wrong weld sequence • tacks too small.

CHAPTER 8

Hardfacing

Chapter contents

The process
Uses
Equipment
Application

The process

Hardfacing is an adaption of the manual metal arc welding process to provide a surface coating for metals.

Uses

Hardfacing provides a wear resistant surface while retaining the original metal's ductility. It has many applications, particularly on equipment which is subject to abrasion through contact with soil and stones.

Equipment

Manual metal arc welding plant as used for normal welding.

Hardfacing electrodes

A large variety of electrodes are available to cope with a wide range of hardnesses and conditions.

Application

Wear conditions

Severe abrasion of earth moving equipment occurs when it is used in soils with a high silicon content, so tools with blades and edges in contact with these will need to be replaced or reclaimed on a regular basis. Also, surfaces subjected to impact need to be hardfaced. If all the tool was hardened it would soon fail in service, because it would not be able to absorb the shock, however, with hardfacing, we can use a more resilient base and protect only those areas actually subject to impact. It is essential to first evaluate the job. Some worn surfaces may need to have the base metal built up before hardfacing, because hardfacing materials are expensive and thick buildups can be costly and unnecessary. In general, hardfacing deposits are never over 5mm thick.

Consumable selection

Hardfacing electrodes depositing a range of carbides are best for abrasion resistance. Electrodes depositing austenitic manganese are best for impact resistance. Electrodes for both abrasion and impact conditions are also available.

Deposition patterns

Large areas of tools in contact with abrasive material are rarely completely covered with hardfacing. Partial and selective coverage with patterns using strings of weld beads will usually be adequate. Some of these patterns are as follows

Parallel bead pattern

The parallel bead pattern allows the part to lift and strike rocks while the base metal is protected. Because the beads are formed in the same direction as the tool moves through or over the earth or rock, there is no restriction to the part's operation (Figure 8.1).

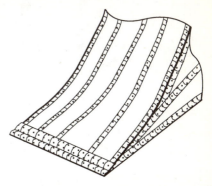

Figure 8.1 *Parallel bead pattern*

Diamond and spot patterns

These patterns provide a degree of self cleaning, because it is more difficult for the dirt to remain deposited between the beads (Figure 8.2).

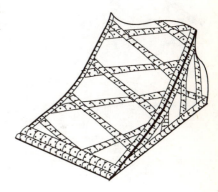

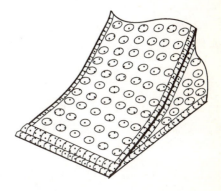

Figure 8.2 *Diamond and spot patterns*

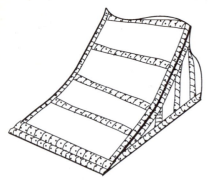

Figure 8.3 *Transverse bead patterns*

Transverse bead pattern

In highly abrasive situations this pattern traps dirt between the beads, protecting the base metal (Figure 8.3).

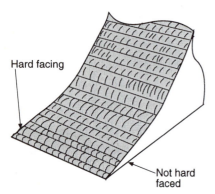

Hard facing

Not hard faced

Figure 8.4 *Self sharpening edge*

Self sharpening edge

By hardfacing only the top part of some earth cutting teeth, a sharp edge can be maintained as the bottom wears down (Figure 8.4).

Preheating

When surfacing one metal with another which is quite different from it, there will be differences in their contraction rates and stress resistances. To prevent the hardfacing material from peeling off, it may be necessary to preheat the part to be hardfaced. This will be most necessary on large parts which, because of their volume, can have a quenching effect on the weld deposits.

Buffer layer

This is a layer applied to the part before the hardfacing is done and it is usually applied to soft metals which are to be hardfaced with a very hard deposit; the buffer layer is of intermediate hardness and adds toughness to the final layer, aiding the prevention of cracking.

CHAPTER 9

Marking out and development

Chapter contents

Basic techniques

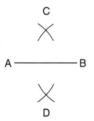
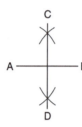

Constructing a centreline at 90° to an existing line.

Using point A as a centre and with a radius greater than half AB, scribe arcs above and below line AB.

With point B as a centre and with the same radius, scribe arcs cutting the previous arcs at C and D.

Join points C and D. The line CD now passes through the centre of line AB at 90°. Figure 9.1.

Figure 9.1

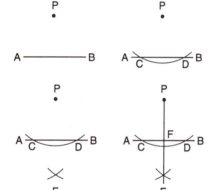

Constructing a line at 90° to a baseline from a given point not on the line.

With centre P, scribe an arc cutting line AB at C and D.

With centres C and D and the same radius, scribe intersecting arcs below the line (point E).

Join points P and E to cut line AB at F.

Line PE is now at 90° to AB. Figure 9.2.

Figure 9.2

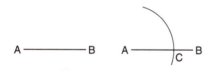

Constructing a line at 60° to a baseline

With centre A and any radius, scribe an arc cutting line AB at C.

With centre C and the same radius, scribe an arc cutting the existing arc at D.

Draw a line from A passing through point D. The angle DAC is 60°. Figure 9.3.

Figure 9.3

To bisect an angle.

With centre B and any radius, scribe an arc cutting lines AB and BC at points D and E respectively.

With centres D and E and any radius, scribe two intersecting arcs at point P.

Join points B and P.

The angle ABP will now be equal to the angle PBC. Figure 9.4.

This method can be used to construct angles of 90°, 45°, 30° and 15° and a combination of these techniques can be used to construct angles of 75°, 135°, 150° and 165°.

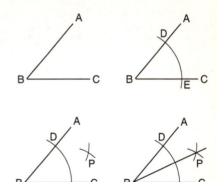

Figure 9.4

Squaring

When large projects are to be marked out, small hand-held squares are very inaccurate. A method of plotting more suitable reference points is needed. Descriptions of some of these methods are as follows.

3:4:5 method

A figure drawn with sides 3 units, 4 units and 5 units long will always form a right angled triangle. It does not matter if these measurements are 3 , 4 and 5 metres or 3 , 4 and 5 kilometres; as long as the ratio 3:4:5 is maintained a right angle will be produced. Figure 9.5.

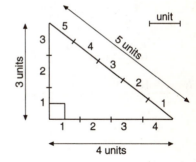

Figure 9.5

Diagonals

The diagonals of a square or a rectangle must always be the same length. If the diagonals are different then the figure is not rectangular, so when constructing square or rectangular frames make sure that the diagonals are the same length, to ensure that the job is rectangular. Figure 9.6.

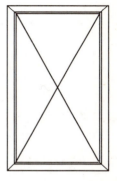

Figure 9.6

Figure 9.7

2 Chords in a semi-circle

Drawing touching chords from each end of a semi-circle will always create a right angle. These chords will meet at 90° no matter where they touch the semi-circle. Figure 9.7.

Formulas

Area of a square $= S \times S$

Figure 9.8

Area of a rectangle $= L \times B$

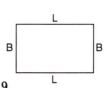

Figure 9.9

Area of a triangle $= \dfrac{\text{Base} \times \text{perpendicular height}}{2}$

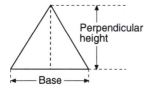

Figure 9.10

Circumference of circle $= \pi D$ or $2\pi R$

where $\pi = 3.1416$

$D =$ diameter

$R =$ radius

Figure 9.11

Area of a circle $= \pi R^2$

where $\pi = 3.1416$

$R =$ radius

Figure 9.12

Relationship between the sides of a right angled triangle

In a right angled triangle the square of the longest side (the hypotenuse) is equal to the sum of the squares of the other two sides.

This is expressed as $a^2 + b^2 = c^2$.

Try it with the 3:4:5 triangle.

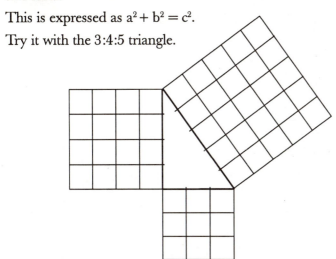

Figure 9.13

This formula is handy for calculating material lengths when only 2 side measurements of a right angled triangle are known.

Sizes of angles in regular figures

The size of each of the angles in any plane figure with equal sides can be easily calculated using the following formula:

$$\frac{(2 \times \text{no. of sides} - 4) \times 90}{\text{no. of sides}}$$

It looks complicated but it is really quite simple; for example, a pentagon:

$$\frac{(2 \times 5 - 4) \times 90}{5}$$

$$= \frac{(10 - 4) \times 90}{5}$$

$$= \frac{6 \times 90}{5}$$

$$= \frac{540}{5}$$

$$= 108$$

Figure 9.14

Each angle in a pentagon is 108°

Development of a rectangular hopper

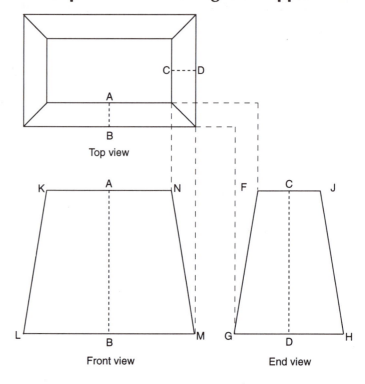

Figure 9.15

Top view

Front view

End view

True shape of front

The line AB in the top view is the centre line of the front of the hopper. The true length of this line is FG in the end view (Figure 9.15).

Step 1

Draw two parallel lines the distance FG apart. Figure 9.16.

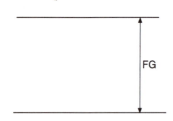

Figure 9.16

Step 2

Draw a centre line intersecting these lines at 90°. Figure 9.17.

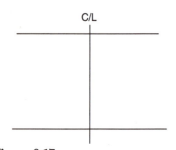

Figure 9.17

Step 3

On the bottom line, each side of the centre line, mark off the distance LB. Figure 9.18.

Figure 9.18

Step 4

On the top line, each side of the centre line, mark off the distance KA. Figure 9.19.

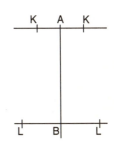

Figure 9.19

Step 5

Join points LK on each side and the true shape of the front is drawn. Figure 9.20.

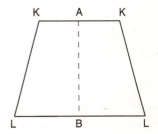

Figure 9.20

True shape of end

The true shape of the end can be drawn in a similar way using length NM, in the front view, as the true length of the centre line. Construct parallel lines at each end of this centre line length. Lengths FC (top) and GD (bottom) can be marked on the parallel lines and the true shape of the end can now be drawn.

The four parts can now be marked out and cut, or they can be marked out, cut and folded, leaving just one seam to weld. Figure 9.21.

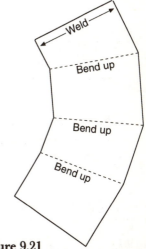

Figure 9.21

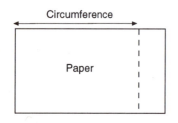

Circumference

Paper

Figure 9.22

Finding the circumference of a pipe

Mathematically

Our first instinct when needing the circumference of a pipe is to measure its diameter and apply the normal formula which is πD. This is accurate but takes time if you do not have a calculator. There is another, easier way.

Paper wrap-around

Simply wrap a piece of paper around the pipe and mark it where the ends overlap. Remove the paper and measure the distance from the end to the mark. Figure 9.22.

To divide a pipe circumference into four equal parts

Mathematically

We can use mathematics and divide πD by 4 or use a paper wrap-around.

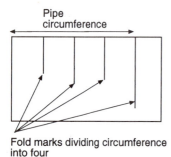

Pipe circumference

Fold marks dividing circumference into four

Figure 9.23

Paper wrap-around

Find the circumference using the wrap-around method. Now carefully fold the paper length in half and fold it again. If you put it back on the pipe you can mark the four divisions directly from the fold marks in the paper. This is a very accurate method. Figure 9.23.

Corner joint in pipe

The 45° angle cuts in pipe to form a 90° joint are difficult to make by guesswork and many people try it with terrible results. The parts require a lot of grinding and the poor fit makes it difficult to weld. There are two ways in which we can mark the pipe accurately. The first involves making a wrap around template after the developed shape has been laid out (Figure 9.24).

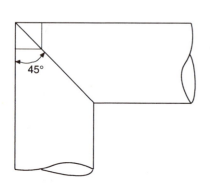

45°

Figure 9.24

Development of a template corner joint in pipe

Step 1

Draw a side view and top view of the pipe shape (the length is unimportant). (Figure 9.25.)

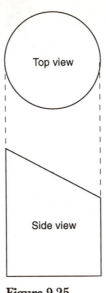

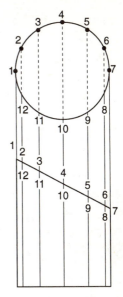

Step 2

Divide the top view into 12 equal parts, number them and project them with vertical lines onto the side view (Figure 9.26).

Figure 9.25   **Figure 9.26**

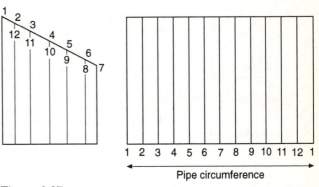

Figure 9.27

Step 3

Find the circumference of the pipe and use this length as a base line for the development. Divide it into 12 equal parts and number them starting and finishing with 1 (Figure 9.27).

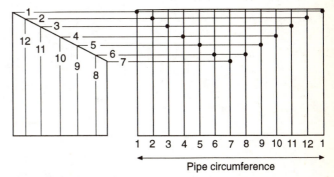

Step 4

Transfer the length of each line from the side view to form a template (Figure 9.28).

Figure 9.28

115

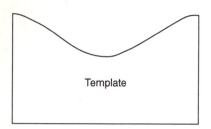

Template

Figure 9.29

Step 5
Join the points with a flowing curve and cut out your wrap around template (Figure 9.29).

Marking a corner joint in pipe using a paper wrap around and a flexicurve.

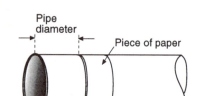

Pipe diameter

Piece of paper

Figure 9.30

Step 1
Use a piece of paper to mark a line around the pipe a distance equal to the diameter of the pipe from its end (Figure 9.30).

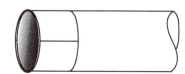

Figure 9.31

Step 2
Mark this section into four equal parts using the folded paper method (Figure 9.31).

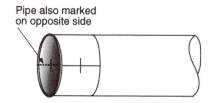

Pipe also marked on opposite side

Figure 9.32

Step 3
Divide two of these lines (opposite each other) in half and mark (use a folded piece of paper) (Figure 9.32).

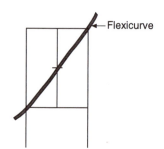

Flexicurve

Figure 9.33

Step 4
Using a flexicurve, mark the cut in two halves by connecting the end of the pipe to the mark furthest along it, making sure the flexicurve passes through one of the middle marks. Mark the shape and repeat this on the other side (Figure 9.33).

A flexicurve is a piece of flexible material encased in rubber. When bent by hand to form a curve it remains so curved until it is rebent. Flexicurves are used by metal fabricators and draftsman to assist in marking out smoothly complicated curves.

Sit-on pipe branch

Developing a template for a sit-on pipe branch (Figure 9.34)

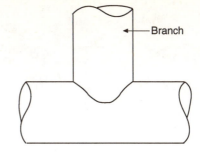

Figure 9.34

Step 1
Mark out a top view and an end view of the joint (Figure 9.35).

Step 2
Divide the top view into 12 equal parts, number them and project them onto the end view (Figure 9.36).

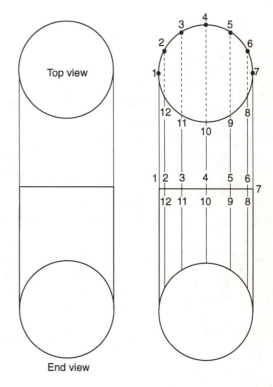

Figure 9.35   **Figure 9.36**

Step 3
Obtain the pipe circumference and use this length as a base line for the development. Divide it into 12 equal parts and number them starting and finishing with 1 (Figure 9.37).

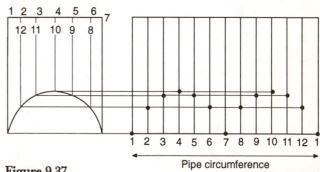

Figure 9.37

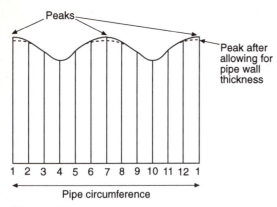

Figure 9.38

Step 4
Transfer the length of each line (from the top of the pipe to the bottom) from Step 2 to the base line (Figure 9.38).

Step 5
Join the marks with a flowing curve then cut it out and wrap it around the pipe and mark. The lengths of the peaks may need to be shortened to allow for the pipe wall thickness.

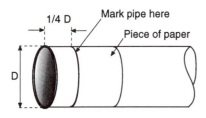

Figure 9.39

Marking a sit-on pipe branch using an off-cut of pipe.

Step 1
Mark a line around the end of the pipe a quarter of its diameter from the end (Figure 9.39).

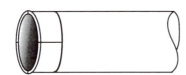

Figure 9.40

Step 2
Divide the circumference of the pipe into four equal sections (Figure 9.40).

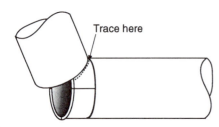

Figure 9.41

Step 3
Place the off-cut over the end of the pipe touching the line (Figure 9.41).

Step 4
Trace around the off-cut.

Figure 9.42

Step 5
Repeat on the other side. The pipe will look like Figure 9.42 and the cut joint is a surprisingly good fit.

Figure 9.43

Radial line development of a conical hopper (Figure 9.43)

Step 1
Draw a side view and top view of the hopper. Divide the top view into 12 sections (Figure 9.44).

Step 2
Find the apex by extending the sides of the hopper. With the apex as a centre scribe 2 arcs (Figure 9.45).

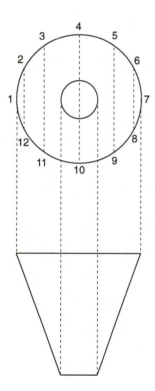

Figure 9.44

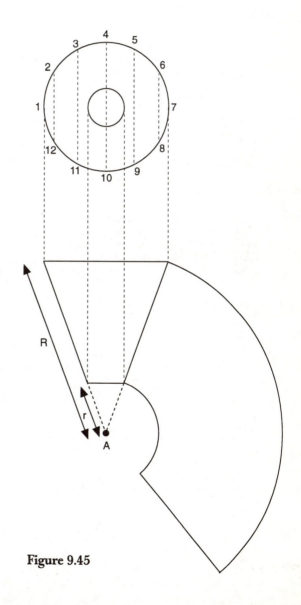

Figure 9.45

Step 3

Step off the large circumference and draw radial lines to apex (Figure 9.46).

Radial lines can be used as a guide when forming the conical shape.

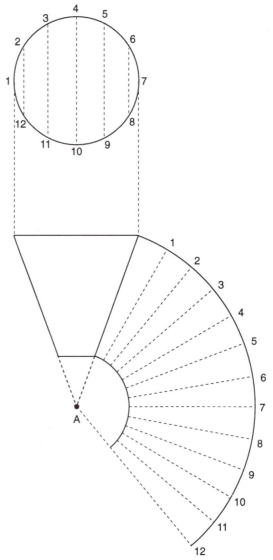

Figure 9.46

CHAPTER 10

Projects

Chapter contents

Jigs
Welded projects
General welding considerations

Jigs

When metal parts need to be assembled or bent to shape, jigs are often used. Jigs ensure that assemblies or bends used more than once on a particular job are the same. The two bending jigs shown here are easy to make and use, and they can be welded using either MMAW or GMAW. However, be sure that you tack all parts of the jigs into their correct positions before fully welding them, because when the parts can be seen in position, but have not yet been fully welded, minor adjustments are easier to make. Good tacking will also help to control distortion during welding.

Two-pin bending jig

This is a simple jig which can be used to bend flat bar (up to 25 mm wide × 6 mm thick) and round bar (up to 8 mm diameter) easily and accurately without heating the steel. Thicker and wider sections will have to be brought to red heat at the bend area. Heat travels quickly through steel so be sure to wear gloves when handling metal pieces which have had some part heated (Figures 10.1 and 10.2).

The pins should be at least 12 mm in diameter and the jig needs to be held in a vice during use.

Figure 10.1 *With a little practice items like gate straps, U-bolts and brackets can made with a two-pin jig*

Material
1 off, 75 x 75 x 10 x 150 mm long mild steel (MS) angle.
2 off, 16 x 50 mm long MS rod.

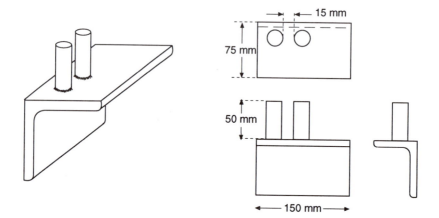

Figure 10.2 *Construction details of the two-pin bending jig*

Radiused bending jig incorporating a bending dog

Heavy metal which cannot be heated is very difficult to bend accurately on a two-pin bending jig, because the leverage necessary causes a curve along the whole piece which has to be hammered out (Figure 10.3) When a jig is made which allows a bending dog to be used, this problem is eliminated.

The jig shown here is, by necessity, a fairly solid construction. Like the two-pin bending jig the metal can be bent either hot or cold, and the jig needs to be held securely in a vice (Figures 10.4 and 10.5).

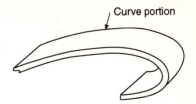

Figure 10.3 *Unacceptable bend owing to cold bending on a two-pin jig*

Material

Jig
1 off, 150 x 75 x 400 mm long steel channel.
1 off, 25 x 25 x 200 mm long steel square bar.
1 off, 75 mm nominal bore (NB) pipe x 65 mm long.
1 off, 25 mm NB pipe x 65 mm long.

Dog
1 off, 25 mm diameter x 550 mm long steel round bar.
2 off, 25 mm diameter x 75 mm long steel round bar.

Note that the handle is off-set to allow its use without reaching across the jig.

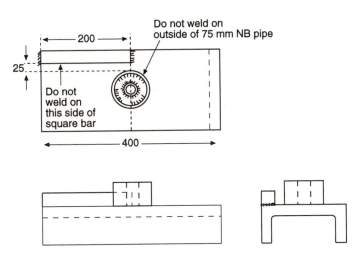

Figure 10.4 *Details of the jig*

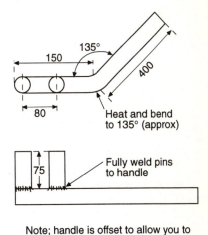

Figure 10.5 *The bending handle*

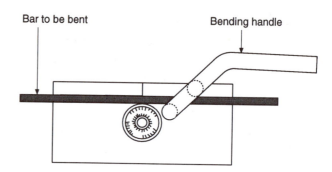

Bar to be bent

Bending handle

Figure 10.6 *The jig in use*

Welded projects

G-clamp

Small G-clamps can be easily made from pieces of scrap plate, which should be at least 12 mm thick. The dimensions shown are a guide only and can be varied to suit the available steel plate (Figure 10.7).

Material

1 off, 12 mm plate 150 x 100 mm.
1 off, 12 mm threaded rod 120 long.
1 off, 6 mm diameter rod 50 mm long.
1 off, 3 mm plate 30 x 20 mm.
1 off, 12 mm nut to suit threaded rod.

The clamp body should be marked out as shown, then carefully flame cut, using guide wheels and a circle cutting attachment (Figure 10.8).

Braze welding can be used to weld the 12 mm nut and the 3 mm bearing plate to the clamp body, and the 6 mm diameter handle to the threaded rod.

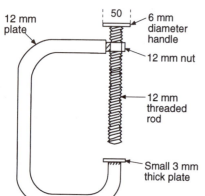

12 mm plate

50

6 mm diameter handle

12 mm nut

12 mm threaded rod

Small 3 mm thick plate

Figure 10.7 *Details of the G-clamp*

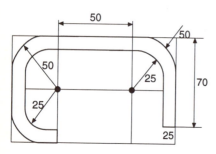

50

50

50

25

70

25

25

25

Figure 10.8 *G-clamp – marking out the body*

Car stand

This is an excellent cutting, assembling and GMA welding exercise. Car stands are useful when extensive repairs are needed on vehicles, because they hold the vehicle in an elevated position much more safely than car jacks. When using car stands, the wheels on the ground should be firmly chocked front and back to prevent the vehicle from moving (Figure 10.9).

Material (for one car stand)

Centre pipe
1 off, 30 mm NB pipe 300 mm long.

Gussets
4 off, 6 mm plate (Figure 10.9).

Base plate
1 off, 6 mm plate 250 x 250 mm.

Top plate
1 off, 30 x 8 flat 100 mm long.

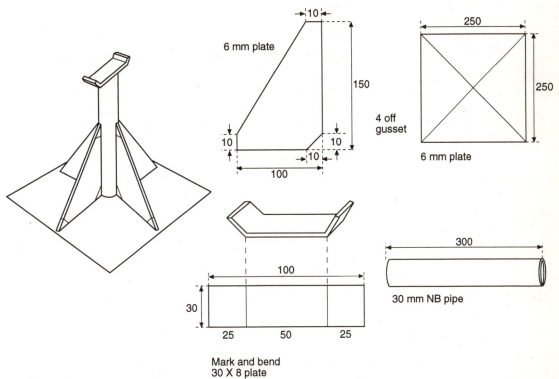

Figure 10.9 *Car stand detail*

Wheel chocks

These will prevent a car from rolling when it is being jacked up or while it has some of its wheels off the ground on car stands. They can be made from 3 mm plate and are easily welded with oxyacetylene. The 3 mm plate can be flame cut using a number 6 or 8 nozzle.

Material (for one chock). See Figure 10.10

Figure 10.10

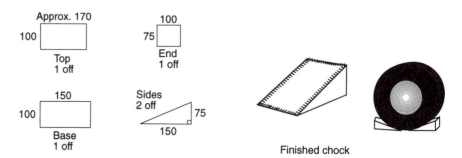

Finished chock

Trolley

Material

Wheels
2 off, 200 mm diameter (12 mm diameter axle hole).

Handles
2 off, 20 mm diameter NB pipe 880 mm long.
2 off, 20 mm diameter NB pipe 200 mm long.

Centre pieces
3 off, 30 x 6 flat x 400 mm long.
2 off, 30 x 6 flat x 270 mm long.

Tray
1 off, 5 mm plate 425 x 150 mm.

Tray gussets
2 off, 5 mm plate.

Axle
1 off, 12 mm rod x 530 mm long.

Axle supports
2 off, 8 mm plate.

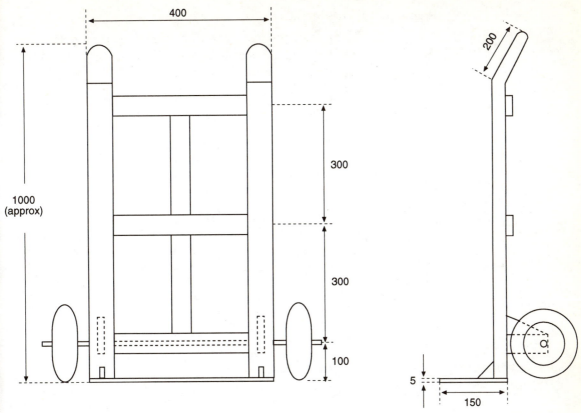

Figure 10.11 *Trolley, general views*

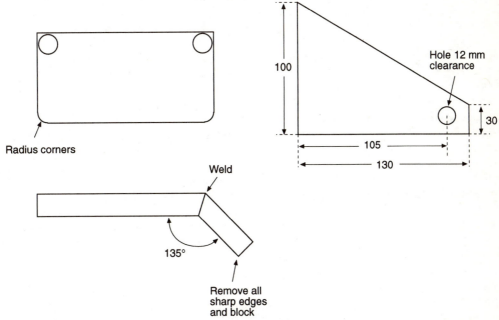

Figure 10.12 *Tray, axle supports and handles*

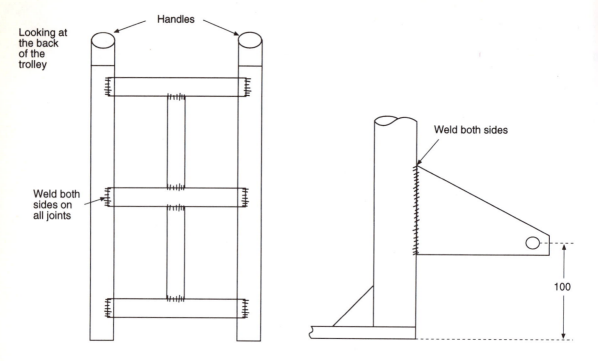

Figure 10.13 *Handles, centre pieces and axle support brackets*

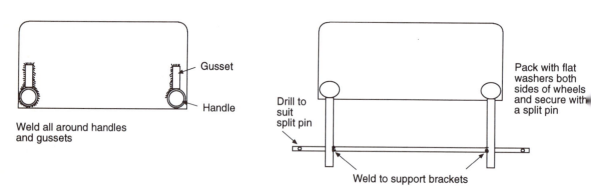

Figure 10.14 *Handles to tray and axles to support brackets*

General welding considerations

The following is a rough guide only, because to provide a complete welding procedure for every probability would be next to impossible. However, this guide will assist you to repair most breakages involving rural machine parts.

Machine part repairs

Breakages to machinery parts and implements can occur at any time. Often the broken parts can be welded, saving down time and the cost of a complete replacement unit. Before welding, the job needs to be carefully assessed as follows:

- Identifying the metal from which the part is made.
- Whether to dismantle the part or not.
- The type of welding process suitable.
- Consumables required.
- Joint preparation.
- Preheating requirements.
- Post-weld treatment.

Type of metal

It is important that you identify the type of metal used for the part:

- Most implement frames and welded structures on tractors and trailers, will be made from low carbon steel, which generally is no problem to weld.
- Parts in contact with the ground will be of a harder and more brittle metal, and will need special attention.
- Parts on stationary machinery, such as hammer mills, may have some of their heavier sections – particularly their bases – made from cast iron, which can be identified by the roughish surface formed by the sand moulds used in the casting process.
- Wearing surfaces, axles, shafts and bearings will generally be of medium to high carbon steel and may require similar attention to those parts used for ground contact such as grader blades.

Dismantling

If the broken parts can be easily realigned in their permanent positions, dismantling may not be necessary. However, all other parts must be protected from welding heat. Also, consideration should be given to preparing the part for welding; for instance, at times the position (vertical or overhead) may not make the job easy to do. In such cases it may be preferable to dismantle the job and prepare and weld the job in the flat position.

Process suitability

Most machinery parts can be welded using GMAW or MMAW. Some jobs, such as breaks in cast iron, or when brazing or braze welding is more suitable, oxyacetylene welding will be necessary. Since this latter process does not need a power supply, the portability of the welding equipment allows for repairs in remote places.

Consumables

Filler rods, electrode wires or covered electrodes must be compatible with the parts being welded.

Joint preparation

All breaks which need repairs using butt joints must be prepared to ensure that welding penetration occurs through the full thickness of the parts.

Broken joints requiring fillet welds are usually caused by fatigue. Strengthening such joints with gussets after rewelding may be necessary to ensure that they do not break again.

Preheating

Medium to high carbon steel, cast iron and low carbon steel over 15 mm thick will all need varying amounts of preheating to prevent rapid cooling, which can cause brittle cracking.

Post weld treatment

Cast iron parts will need to be reheated to an all-over even temperature, while joints in other metals should be allowed to cool slowly in still air. On no occasion should any welded joint be immersed in water or hosed to cool it down. This will alter the metallurgical structure of the metal and can lead to cracking in or alongside the weld.

Hard facing

Seek advice from your welding products suppliers. Remember that the source of wear on the part (impact, abrasion, heat or a combination of these) will dictate the type of surfacing consumables you will need.

Table 10.1 *Summary of conditions and processes*

Type of metal	Low carbon steel	Medium to high carbon steel	Cast iron	Stainless steel
Identification	Used on most welded frames Will usually distort (bend) before breaking	Breaks sharply without distortion Hard, usually with very smooth surfaces and sharp edges	Used on heavy or intricate shapes Will have some rough surfaces Very soft to file	Hard and shiny Free of rust Usually non-magnetic
Process	Oxyacetylene GMAW MMAW	GMAW MMAW	Oxyacetylene	GMAW MMAW
Consumables	Steel filler rod Steel GMAW wire General purpose covered electrodes e.g. E4113	Steel GMAW wire (these are hydrogen controlled) Hydrogen controlled covered electrodes e.g.E4816, E4818	Cast iron filler rod Cast iron flux	Stainless steel GMAW wire Stainless steel covered electrodes
Preheating	Do not weld when wet or iced over. Preheat till warm to touch Metal over 15 mm thick should also be preheated to 50°C	Must be warm to touch before welding Preheat to 50° to 100°C	Preheat all over to a dull red before welding Reheat evenly	Usually does not require preheating unless over 15 mm thick
Post-weld treatment	Post heating not generally required	Post heat may be required on thick sections Cool slowly	Cool slowly in warm sand	Post heat not generally required

Index

Some other titles in the
PRACTICAL FARMING SERIES
Published by Inkata

BEEF CATTLE: Breeding, Feeding and Showing — Lucy Newham

CROP SPRAYING: Techniques and Equipment — Gary Alcorn

FARM WATER SUPPLIES: Planning and Installation — Neil Southorn

FENCES AND GATES: Design and Construction — David East

FIRE FIGHTING: Management and Techniques — Frank Overton

PASTURE MANAGEMENT — Rick Bickford

RISK MANAGEMENT: Rural Property Planning — Mike Krause

RURAL SAFETY: Stock, Machinery and General Hazards — Andrew Brown & Brian Lawler

RURAL SAFETY: Chemicals and Dangerous Substances — Andrew Brown, Brian Lawler & David Smith

SMALL PETROL ENGINES: Operation and Maintenance — Bruce Holt

TRACTORS: Operation, Performance and Maintenance — Neil Southorn

WORKING DOGS: Training for Sheep and Cattle — Colin Seis